Learn

Eureka Math®

Grade 2

Modules 6 & 7

Published by Great Minds®.

Printed in the U.S.A.
This book may be purchased from the publisher at eureka-math.org.
BAB 10 9 8 7 6 5 4 3 2

ISBN 978-1-64054-057-6

G2-M6-M7-L-05.2018

Learn • Practice • Succeed

Eureka Math® student materials for *A Story of Units*® (K–5) are available in the *Learn, Practice, Succeed* trio. This series supports differentiation and remediation while keeping student materials organized and accessible. Educators will find that the *Learn, Practice,* and *Succeed* series also offers coherent—and therefore, more effective—resources for Response to Intervention (RTI), extra practice, and summer learning.

Learn

Eureka Math Learn serves as a student's in-class companion where they show their thinking, share what they know, and watch their knowledge build every day. *Learn* assembles the daily classwork—Application Problems, Exit Tickets, Problem Sets, templates—in an easily stored and navigated volume.

Practice

Each *Eureka Math* lesson begins with a series of energetic, joyous fluency activities, including those found in *Eureka Math Practice.* Students who are fluent in their math facts can master more material more deeply. With *Practice,* students build competence in newly acquired skills and reinforce previous learning in preparation for the next lesson.

Together, *Learn* and *Practice* provide all the print materials students will use for their core math instruction.

Succeed

Eureka Math Succeed enables students to work individually toward mastery. These additional problem sets align lesson by lesson with classroom instruction, making them ideal for use as homework or extra practice. Each problem set is accompanied by a Homework Helper, a set of worked examples that illustrate how to solve similar problems.

Teachers and tutors can use *Succeed* books from prior grade levels as curriculum-consistent tools for filling gaps in foundational knowledge. Students will thrive and progress more quickly as familiar models facilitate connections to their current grade-level content.

Students, families, and educators:

Thank you for being part of the *Eureka Math*® community, where we celebrate the joy, wonder, and thrill of mathematics.

In the *Eureka Math* classroom, new learning is activated through rich experiences and dialogue. The *Learn* book puts in each student's hands the prompts and problem sequences they need to express and consolidate their learning in class.

What is in the Learn book?

Application Problems: Problem solving in a real-world context is a daily part of *Eureka Math.* Students build confidence and perseverance as they apply their knowledge in new and varied situations. The curriculum encourages students to use the RDW process—Read the problem, Draw to make sense of the problem, and Write an equation and a solution. Teachers facilitate as students share their work and explain their solution strategies to one another.

Problem Sets: A carefully sequenced Problem Set provides an in-class opportunity for independent work, with multiple entry points for differentiation. Teachers can use the Preparation and Customization process to select "Must Do" problems for each student. Some students will complete more problems than others; what is important is that all students have a 10-minute period to immediately exercise what they've learned, with light support from their teacher.

Students bring the Problem Set with them to the culminating point of each lesson: the Student Debrief. Here, students reflect with their peers and their teacher, articulating and consolidating what they wondered, noticed, and learned that day.

Exit Tickets: Students show their teacher what they know through their work on the daily Exit Ticket. This check for understanding provides the teacher with valuable real-time evidence of the efficacy of that day's instruction, giving critical insight into where to focus next.

Templates: From time to time, the Application Problem, Problem Set, or other classroom activity requires that students have their own copy of a picture, reusable model, or data set. Each of these templates is provided with the first lesson that requires it.

Where can I learn more about Eureka Math *resources?*

The Great Minds® team is committed to supporting students, families, and educators with an ever-growing library of resources, available at eureka-math.org. The website also offers inspiring stories of success in the *Eureka Math* community. Share your insights and accomplishments with fellow users by becoming a *Eureka Math* Champion.

Best wishes for a year filled with aha moments!

Jill Diniz

Jill Diniz
Director of Mathematics
Great Minds

The Read–Draw–Write Process

The *Eureka Math* curriculum supports students as they problem-solve by using a simple, repeatable process introduced by the teacher. The Read–Draw–Write (RDW) process calls for students to

1. Read the problem.
2. Draw and label.
3. Write an equation.
4. Write a word sentence (statement).

Educators are encouraged to scaffold the process by interjecting questions such as

- What do you see?
- Can you draw something?
- What conclusions can you make from your drawing?

The more students participate in reasoning through problems with this systematic, open approach, the more they internalize the thought process and apply it instinctively for years to come.

Contents

Module 6: Foundations of Multiplication and Division

Topic A: Formation of Equal Groups

Topic B: Arrays and Equal Groups

Topic C: Rectangular Arrays as a Foundation for Multiplication and Division

Topic D: The Meaning of Even and Odd Numbers

Module 7: Problem Solving with Length, Money, and Data

Grade 2
Module 6

Julisa has 12 stuffed animals. She wants to put the same number of animals in each of her 3 baskets.

a. Draw a picture to show how she can put the animals into 3 equal groups.

b. Complete the sentence.

Julisa put ______ animals in each basket.

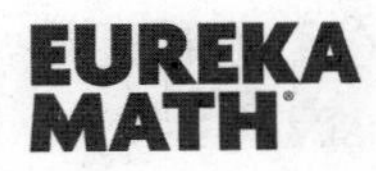

Name ______________________________ Date ______________

1. Circle groups of two apples.

There are _____ groups of two apples.

2. Circle groups of three balls.

There are _____ groups of three balls.

3. Redraw the 12 oranges into 4 equal groups.

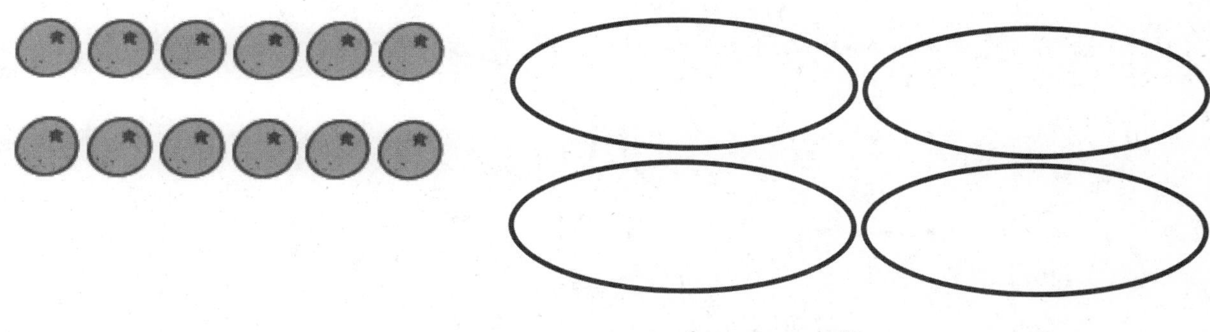

4 groups of _____ oranges

4. Redraw the 12 oranges into 3 equal groups.

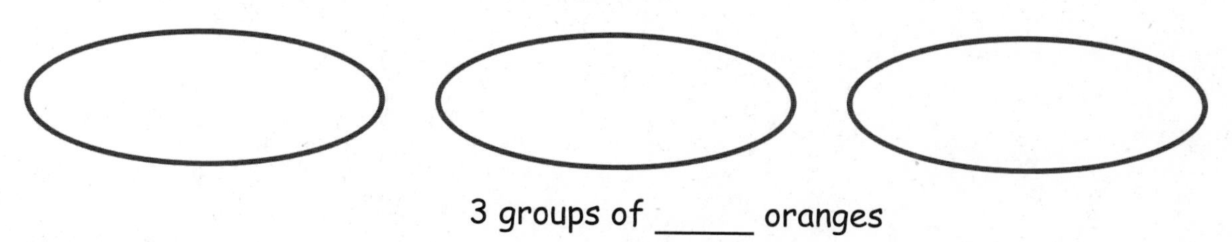

3 groups of _____ oranges

5. Redraw the flowers to make each of the 3 groups have an equal number.

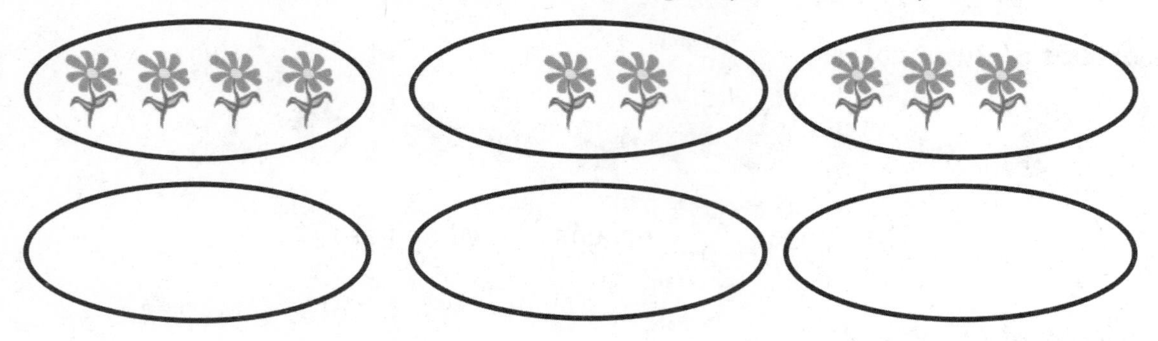

3 groups of ______ flowers = _____ flowers.

6. Redraw the lemons to make 2 equal size groups.

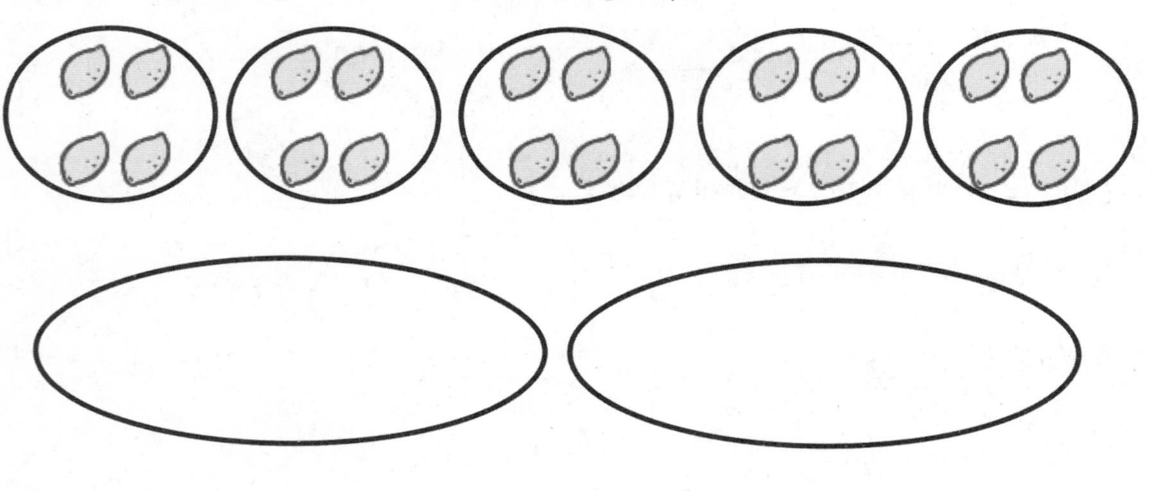

2 groups of ______ lemons = _____ lemons.

Name ______________________________ Date ______________

1. Circle groups of 4 hats.

2. Redraw the smiley faces into 2 equal groups.

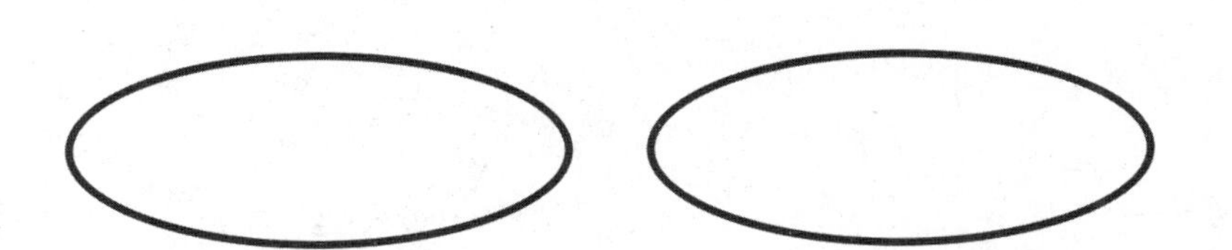

2 groups of _____ = _____.

Mayra sorts her socks by color. She has 4 purple socks, 4 yellow socks, 4 pink socks, and 4 orange socks.

a. Draw groups to show how Mayra sorts her socks.

b. Write a repeated addition equation to match.

c. How many socks does Mayra have in all?

Name ______________________________ Date ________________

1. Write a repeated addition equation to show the number of objects in each group. Then, find the total.

a.

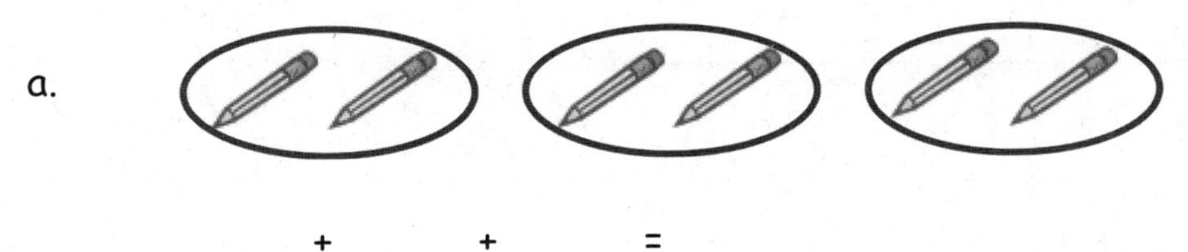

_____ + _____ + _____ = _____

3 groups of _____ = _____

b.

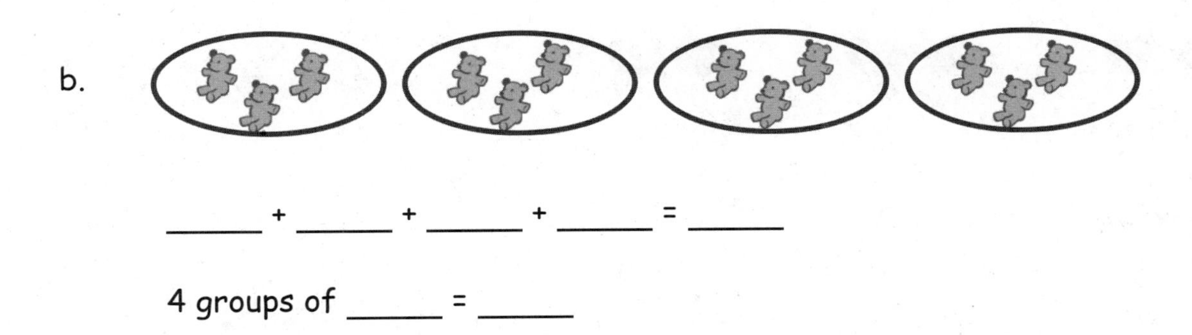

_____ + _____ + _____ + _____ = _____

4 groups of _____ = _____

2. Draw 1 more group of four. Then, write a repeated addition equation to match.

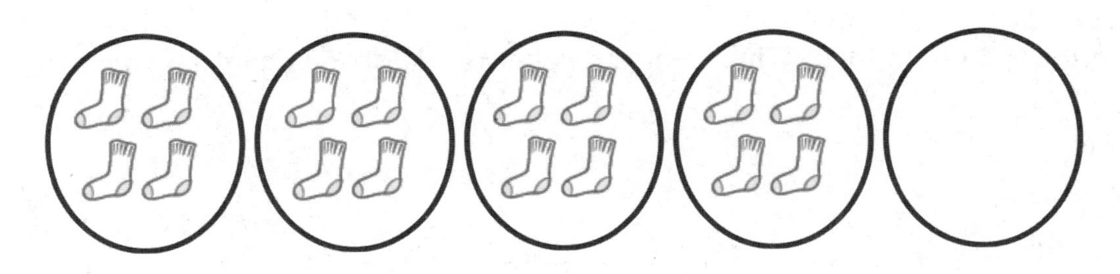

_____ + _____ + _____ + _____ + _____ = _____

5 groups of _____ = _____

3. Draw 1 more group of three. Then, write a repeated addition equation to match.

_____ + _____ + _____ + _____ = _____

_____ groups of 3 = _____

4. Draw 2 more equal groups. Then, write a repeated addition equation to match.

_____ + _____ + _____ + _____ + _____ = _____

_____ groups of 2 = _____

5. Draw 3 groups of 5 stars. Then, write a repeated addition equation to match.

Name ______________________________ Date ______________

1. Draw 1 more equal group.

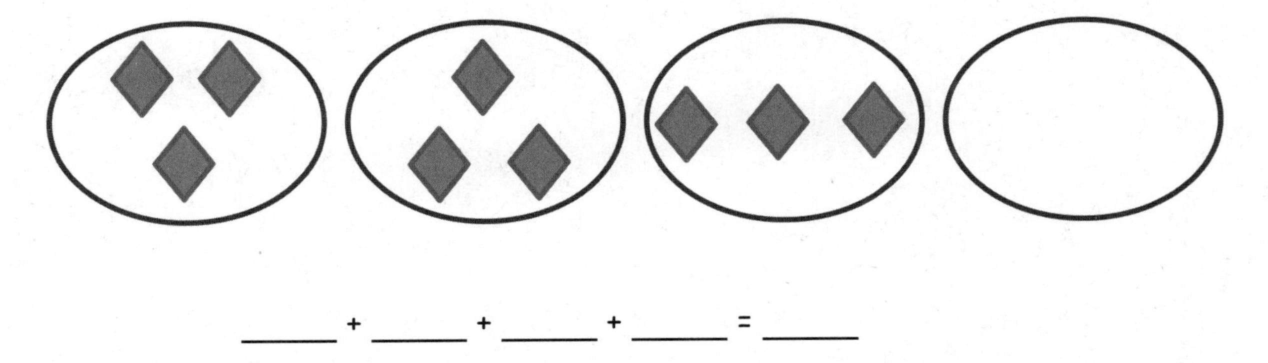

_____ + _____ + _____ + _____ = _____

4 groups of _____ = _____

2. Draw 2 groups of 3 stars. Then, write a repeated addition equation to match.

Markers come in packs of 2. If Jessie has 6 packs of markers, how many markers does she have in all?

a. Draw groups to show Jessie's packs of markers.

b. Write a repeated addition equation to match your drawing.

c. Group addends into pairs, and add to find the total.

Name ______________________________ Date ________________

1. Write a repeated addition equation to match the picture. Then, group the addends into pairs to show a more efficient way to add.

a.

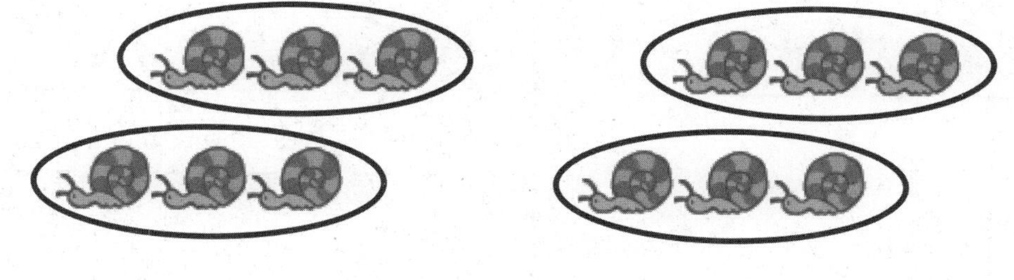
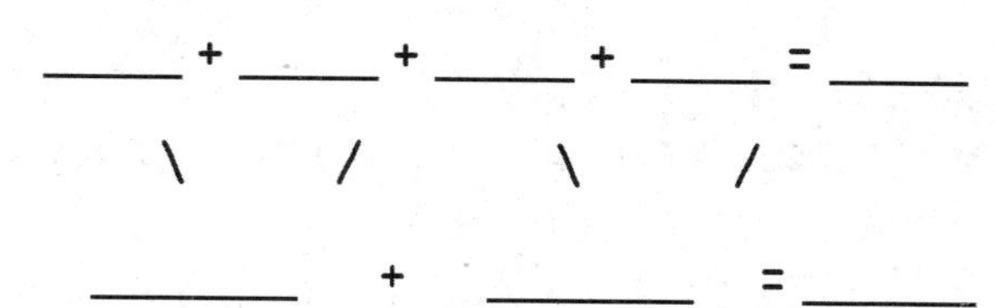

_____ + _____ + _____ + _____ = _____

\ / \ /

_______ + _______ = ______

4 groups of _______ = 2 groups of _______

b.

_____ + _____ + _____ + _____ = _____

_____ + _____ = _____

4 groups of ____ = 2 groups of ____

c.

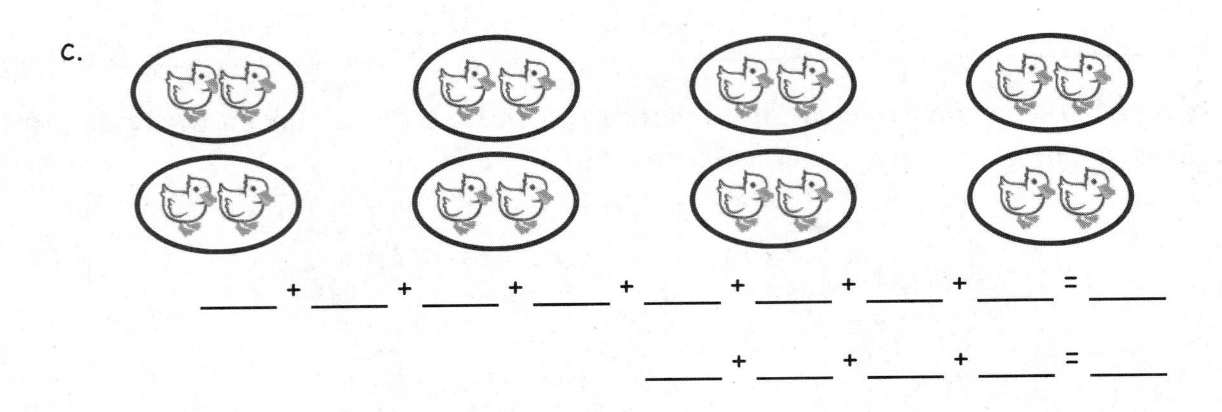

____ + ____ + ____ + ____ + ____ + ____ + ____ + ____ = ____

____ + ____ + ____ + ____ = ____

8 groups of ____ = 4 groups of ____

2. Write a repeated addition equation to match the picture. Then, group addends into pairs, and add to find the total.

a.

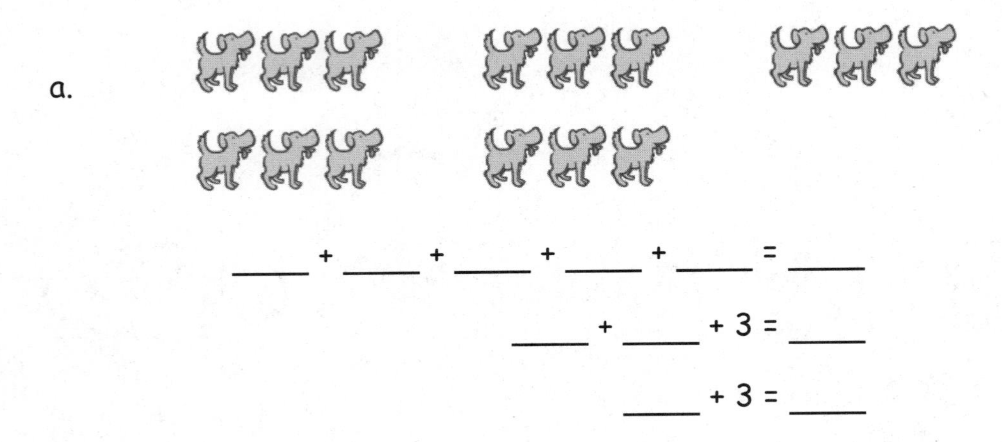

____ + ____ + ____ + ____ + ____ = ____

____ + ____ + 3 = ____

____ + 3 = ____

b.

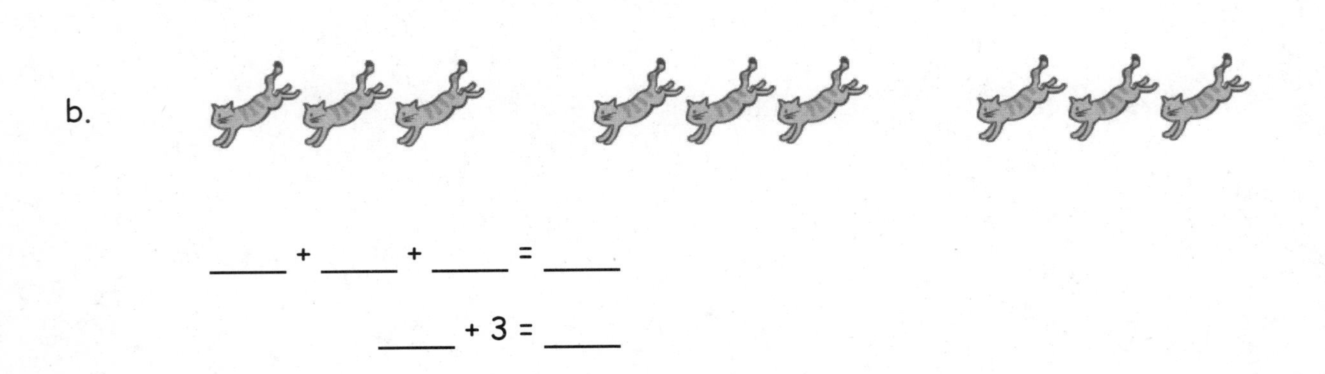

____ + ____ + ____ = ____

____ + 3 = ____

Name ____________________________ Date ______________

Write a repeated addition equation to match the picture. Then, group the addends into pairs to show a more efficient way to add.

_____ + _____ + _____ + _____ = _____

_____ + _____ = _____

4 groups of _____ = 2 groups of _____

R (Read the problem carefully.)

The flowers are blooming in Maria's garden. There are 3 roses, 3 buttercups, 3 sunflowers, 3 daisies, and 3 tulips. How many flowers are there in all?

a. Draw a tape diagram to match the problem.

b. Write a repeated addition equation to solve.

W (Write a Statement that matches the story.)

Name ______________________________ Date ______________

1. Write a repeated addition equation to find the total of each tape diagram.

a. 

_____ + _____ + _____ + _____ = _____

4 groups of 2 = _____

b. 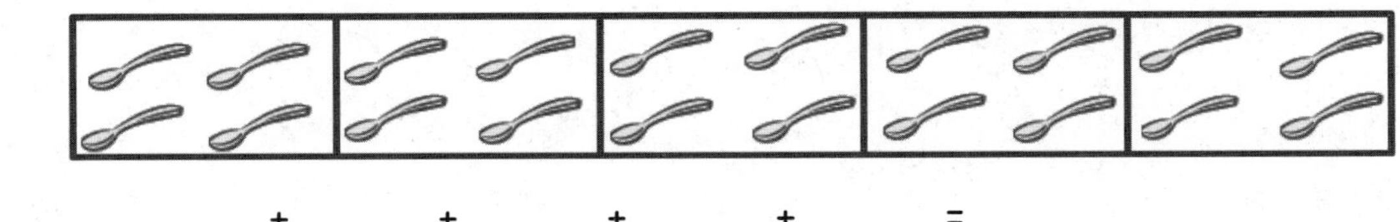

_____ + _____ + _____ + _____ + _____ = _____

5 groups of _____ = _____

c. 

_____ + _____ + _____ = _____

3 groups of _____ = _____

d. 

_____ + _____ + _____ + _____ + _____ + _____ = _____

_____ groups of _____ = _____

2. Draw a tape diagram to find the total.

   a. 3 + 3 + 3 + 3 = _____

   b. 4 + 4 + 4 = _____

   c. 5 groups of 2

   d. 4 groups of 4

   e.

Name ________________________________ Date ________________

Draw a tape diagram to find the total.

1. 

2. 3 groups of 3

3. 2 + 2 + 2 + 2 + 2

Mrs. White is in line at the bank. There are 4 teller windows, and 3 people are standing in line at each window.

a. Draw an array to show the people in line at the bank.

b. Write the total number of people.

___

___

___

Name ____________________________ Date ______________

1. Circle groups of four. Then, draw the triangles into 2 equal rows.

2. Circle groups of two. Redraw the groups of two as rows and then as columns.

3. Circle groups of three. Redraw the groups of three as rows and then as columns.

4. Count the objects in the arrays from left to right by rows and by columns. As you count, circle the rows and then the columns.

a. 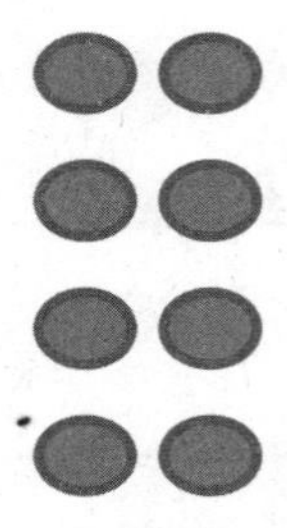

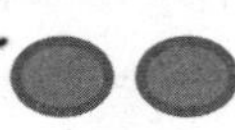

b. 

5. Redraw the circles and stars in Problem 4 as columns of two.

6. Draw an array with 15 triangles.

7. Show a different array with 15 triangles.

Name ______________________________ Date ______________

1. Circle groups of three. Redraw the groups of three as rows and then as columns.

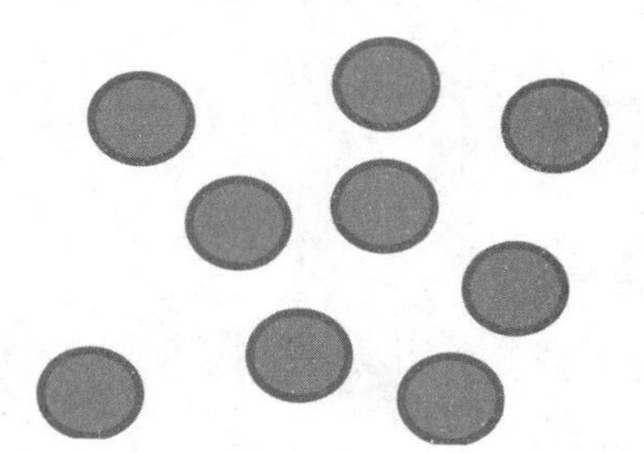

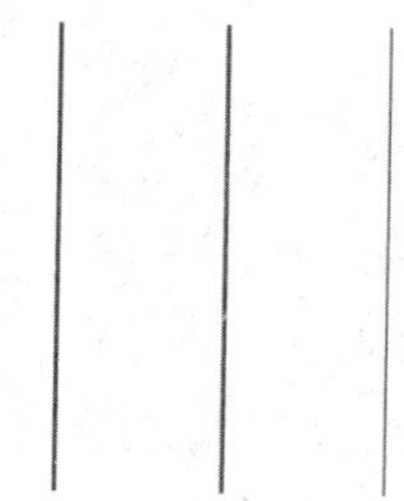

2. Complete the array by drawing more triangles. The array should have 12 triangles in all.

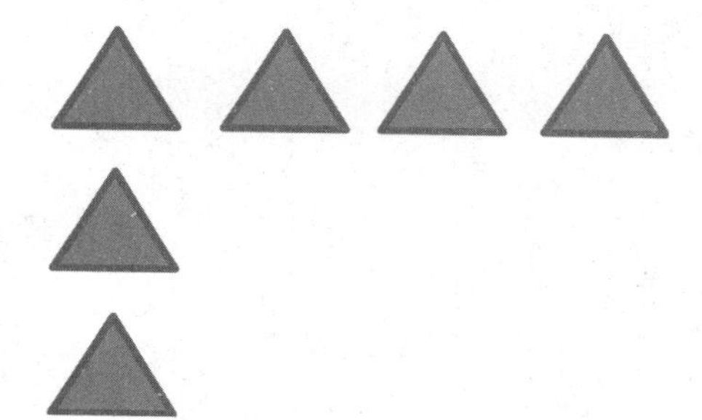

Sam is organizing her greeting cards. She has 8 red cards and 8 blue cards. She puts the red cards in 2 columns and the blue ones in 2 columns to make an array.

a. Draw a picture of Sam's greeting cards in the array.

b. Write a statement about Sam's array.

Name ______________________________ Date ______________

1. Complete each missing part describing each array.

Circle rows.

a.

5 rows of ______ = ______

___ + ___ + ___ + ___ + ___ = ____

Circle columns.

b.

3 columns of ______ = ______

_____ + _____ + _____ = _____

Circle rows.

c.

4 rows of ______ = ______

___ + ___ + ___ + ___ = ___

Circle columns.

d.

5 columns of ______ = ______

___ + ___ + ___ + ___ + ___ = ___

2. Use the array of triangles to answer the questions below.

 a. ____ rows of ____ = 12

 b. ____ columns of ____ = 12

 c. ______ + ______ + ______ = ______

 d. Add 1 more row. How many triangles are there now? ______

 e. Add 1 more column to the new array you made in 2(d). How many triangles are there now? ______

3. Use the array of squares to answer the questions below.

 a. ______ + ______ + ______ + ______ + ______ = ______

 b. ____ rows of ____ = ____

 c. ____ columns of ____ = ____

 d. Remove 1 row. How many squares are there now? ______

 e. Remove 1 column from the new array you made in 3(d). How many squares are there now? ______

Name ______________________________ Date ______________

Use the array to answer the questions below.

a. ____ rows of ____ = _____

b. ____ columns of ____ = _____

c. _____ + _____ + _____ + _____ = _____

d. Add 1 more row. How many stars are there now? _____

e. Add 1 more column to the new array you made in (d). How many stars are there now? _____

R (Read the problem carefully.)

Bobby puts 3 rows of tile in his kitchen to make a design. He lays 5 tiles in each row.

a. Draw a picture of Bobby's tiles.

b. Write a repeated addition equation to solve for the total number of tiles Bobby used.

W (Write a statement that matches the story.)

Name ______________________________ Date ______________

1. a. One row of an array is drawn below. Complete the array with X's to make 3 rows of 4. Draw horizontal lines to separate the rows.

X X X X

b. Draw an array with X's that has 3 columns of 4. Draw vertical lines to separate the columns. Fill in the blanks.

____ + ____ + ____ = ____

3 rows of 4 = ____

3 columns of 4 = ____

2. a. Draw an array of X's with 5 columns of three.

b. Draw an array of X's with 5 rows of three. Fill in the blanks below.

____ + ____ + ____ + ____ + ____ = ____

5 columns of three = ____

5 rows of three = ____

In the following problems, separate the rows or columns with horizontal or vertical lines.

3. Draw an array of X's with 4 rows of 3.

______ + ______ + ______ + ______ = ______

4 rows of 3 = ______

4. Draw an array of X's with 1 more row of 3 than the array in Problem 3. Write a repeated addition equation to find the total number of X's.

5. Draw an array of X's with 1 less column of 5 than the array in Problem 4. Write a repeated addition equation to find the total number of X's.

Name ______________________________ Date ______________

Use horizontal or vertical lines to separate the rows or columns.

1. Draw an array of X's with 3 rows of 5.

___ + ___ + ___ = ___

3 rows of 5 = _____

2. Draw an array of X's with 1 more row than the above array. Write a repeated addition equation to find the total number of X's.

Charlie has 16 blocks in his room. He wants to build equal towers with 5 blocks each.

a. Draw a picture of Charlie's towers.

b. How many towers can Charlie make?

c. How many more blocks does Charlie need to make equal towers of 5?

Name ________________________________ Date ______________

1. Create an array with the squares.

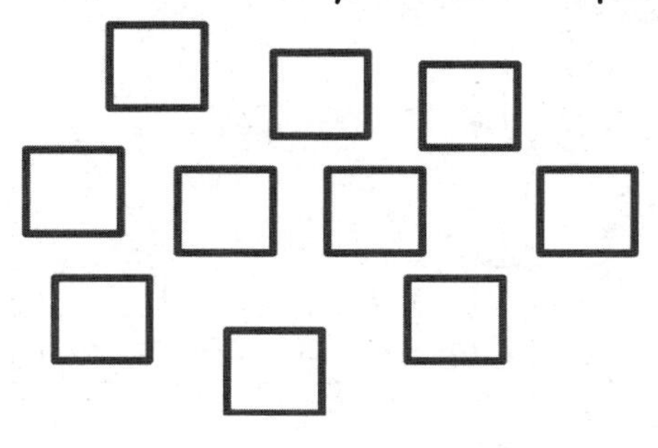

____

____

2. Create an array with the squares from the set above.

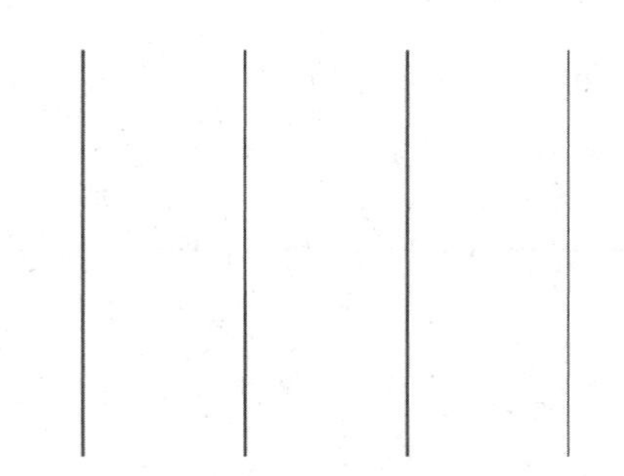

3. Use the array of squares to answer the questions below.

a. There are _____ squares in each row.

b. _____ + _____ = _____

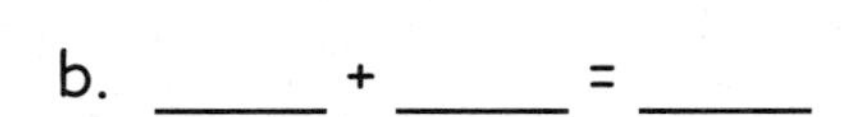

c. There are _____ squares in each column.

d. _____ + _____ + _____ + _____ + _____ = _____

4. Use the array of squares to answer the questions below.

a. There are _____ squares in one row.

b. There are _____ squares in one column.

c. _____ + _____ + _____ = _____

d. 3 columns of _____ = _____ rows of _____ = _____ total

5. a. Draw an array with 8 squares that has 2 squares in each column.

b. Write a repeated addition equation to match the array.

6. a. Draw an array with 20 squares that has 4 squares in each column.

b. Write a repeated addition equation to match the array.

c. Draw a tape diagram to match your repeated addition equation and array.

Name ______________________________ Date ______________

1. Use the array of squares to answer the questions below.

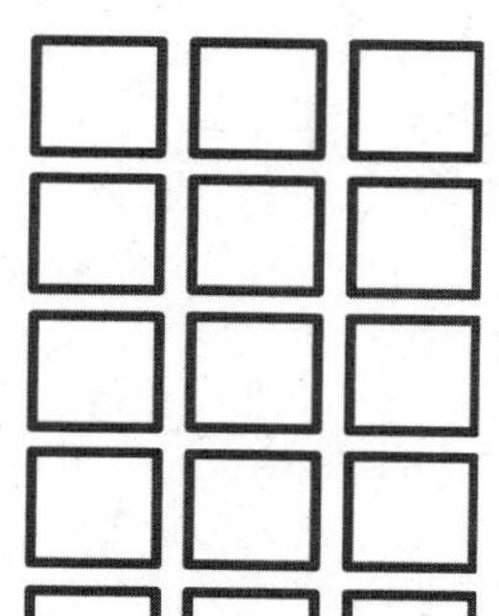

a. There are ____ squares in one row.

b. There are ____ squares in one column.

c. ____ + ____ + ____ = ____

d. 3 columns of ____ = ____ rows of ____ = ____ total

2. a. Draw an array with 10 squares that has 5 squares in each column.

b. Write a repeated addition equation to match the array.

Name ____________________ Date ____________

Draw an array for each word problem. Write a repeated addition equation to match each array.

1. Jason collected some rocks. He put them in 5 rows with 3 stones in each row. How many stones did Jason have altogether?

2. Abby made 3 rows of 4 chairs. How many chairs did Abby use?

3. There are 3 wires and 5 birds sitting on each of them. How many birds in all are on the wires?

4. Henry's house has 2 floors. There are 4 windows on each floor that face the street. How many windows face the street?

Draw a tape diagram for each word problem. Write a repeated addition equation to match each tape diagram.

5. Each of Maria's 4 friends has 5 markers. How many markers do Maria's friends have in all?

6. Maria also has 5 markers. How many markers do Maria and her friends have in all?

Draw a tape diagram and an array. Then, write a repeated addition equation to match.

7. In a card game, 3 players get 4 cards each. One more player joins the game. How many total cards should be dealt now?

Name ______________________________ Date ______________

Draw a tape diagram or an array for each word problem. Then, write a repeated addition equation to match.

1. Joshua cleans 3 cars every hour at work. He worked 4 hours on Saturday. How many cars did Joshua clean on Saturday?

2. Olivia put 5 stickers on each page in her sticker album. She filled 5 pages with stickers. How many stickers did Olivia use?

R (Read the problem carefully.)

Sandy's toy telephone has buttons arranged in 3 columns and 4 rows.

a. Draw a picture of Sandy's telephone.

b. Write a repeated addition equation to show the total number of buttons on Sandy's telephone.

W (Write a statement that matches the story.)

____

____

____

Name ______________________________ Date ______________

Use your square tiles to construct the following rectangles with no gaps or overlaps. Write a repeated addition equation to match each construction.

1. a. Construct a rectangle with 2 rows of 3 tiles.

   ____

   b. Construct a rectangle with 2 columns of 3 tiles.

   ____

2. a. Construct a rectangle with 5 rows of 2 tiles.

   ____

   b. Construct a rectangle with 5 columns of 2 tiles.

   ____

3. a. Construct a rectangle of 9 tiles that has equal rows and columns.

____

b. Construct a rectangle of 16 tiles that has equal rows and columns.

____

4. a. What shape is the array pictured below? ____________________________

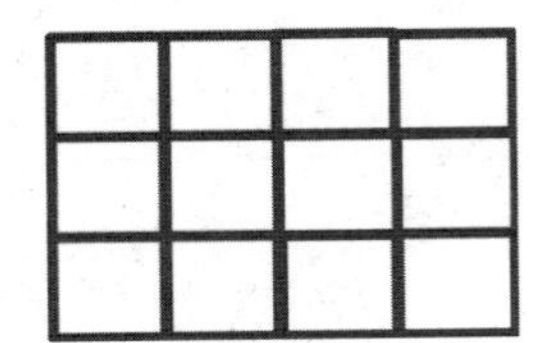

b. Redraw the above shape with one column removed in the space below.

c. What shape is the array now? ____________________________

Name ________________________________ Date ________________

On this sheet, use your square tiles to construct the following arrays with no gaps or overlaps on this sheet. Write a repeated addition equation to match your construction.

1. a. Construct a rectangle with 2 rows of 5 tiles.

   b. Write the repeated addition equation. ____________________________

2. a. Construct a rectangle with 5 columns of 2 tiles.

   b. Write the repeated addition equation. ____________________________

Ty bakes two pans of brownies. In the first pan, he cuts 2 rows of 8. In the second pan, he cuts 4 rows of 4.

a. Draw a picture of Ty's brownie pans.

b. Write a repeated addition equation to show the total number of brownies in each pan.

c. How many brownies did Ty bake altogether? Write an equation and a statement to show your answer.

____

____

____

Name ______________________________ Date ______________

Use your square tiles to construct the following arrays with no gaps or overlaps. Write a repeated addition equation to match each construction.

1. a. Place 8 square tiles in a row.

   b. Construct an array with the 8 square tiles.

   c. Write a repeated addition equation to match the new array.

   ____

2. a. Construct an array with 12 squares.

   b. Write a repeated addition equation to match the array.

   ____

   c. Rearrange the 12 squares into a different array.

   d. Write a repeated addition equation to match the new array.

   ____

3. a. Construct an array with 20 squares.

   b. Write a repeated addition equation to match the array.

   ____

   c. Rearrange the 20 squares into a different array.

   d. Write a repeated addition equation to match the new array.

   ____

4. Construct 2 arrays with 6 squares.

   a. 2 rows of ______ = ______

   b. 3 rows of ______ = 2 rows of ______

5. Construct 2 arrays with 10 squares.

   a. 2 rows of ______ = ______

   b. 5 rows of ______ = 2 rows of ______

Name ______________________________ Date ______________

a. Construct an array with 12 square tiles.

b. Write a repeated addition equation to match the array.

____

Lulu made a pan of brownies. She cut them into 3 rows and 3 columns.

a. Draw a picture of Lulu's brownies in the pan.

b. Write a number sentence to show how many brownies Lulu has.

c. Write a statement about Lulu's brownies.

Extension: How should Lulu cut her brownies if she wants to equally serve 12 people? 16 people? 20 people?

___

___

___

Name ______________________________ Date ______________

1. Draw without using a square tile to make an array with 2 rows of 5.

2 rows of 5 = ______

_____ + _____ = _____

2. Draw without using a square tile to make an array with 4 columns of 3.

4 columns of 3 = ______

_____ + _____ + _____ + _____ = _____

3. Complete the following arrays without gaps or overlaps. The first tile has been drawn for you.

   a. 3 rows of 4

   b. 5 columns of 3

   c. 5 columns of 4

Name ______________________________ Date ______________

Draw an array of 3 columns of 3 starting with the square below without gaps or overlaps.

□

Ellie bakes a square pan of lemon bars, which she cut into nine equal pieces. Her brothers eat 1 row of her treats. Then, her mom eats 1 column.

a. Draw a picture of Ellie's lemon bars before any are eaten. Write a number sentence to show how to find the total.

b. Write an X on the bars that her brothers eat. Write a new number sentence to show how many are left.

c. Draw a line through the bars that her mom eats. Write a new number sentence to show how many are left.

d. How many bars are left? Write a statement.

Name ______________________________ Date ______________

Use your square tiles to complete the steps for each problem.

Problem 1

Step 1: Construct a rectangle with 4 columns of 3.

Step 2: Separate 2 columns of 3.

Step 3: Write a number bond to show the whole and two parts. Then, write a repeated addition sentence to match each part of the number bond.

Problem 2

Step 1: Construct a rectangle with 5 rows of 2.

Step 2: Separate 2 rows of 2.

Step 3: Write a number bond to show the whole and two parts. Write a repeated addition sentence to match each part of the number bond.

Problem 3

Step 1: Construct a rectangle with 5 columns of 3.

Step 2: Separate 3 columns of 3.

Step 3: Write a number bond to show the whole and two parts. Write a repeated addition sentence to match each part of the number bond.

4. Use 12 square tiles to construct a rectangle with 3 rows.

   a. ______ rows of ______ = 12

   b. Remove 1 row. How many squares are there now? ______

   c. Remove 1 column from the new rectangle you made in 4(b). How many squares are there now? ______

5. Use 20 square tiles to construct a rectangle.

   a. ______ rows of ______ = ______

   b. Remove 1 row. How many squares are there now? ______

   c. Remove 1 column from the new rectangle you made in 5(b). How many squares are there now? ______

6. Use 16 square tiles to construct a rectangle.

   a. ______ rows of ______ = ______

   b. Remove 1 row. How many squares are there now? ______

   c. Remove 1 column from the new rectangle you made in 6(b). How many squares are there now? ______

Name ______________________________ Date ________________

Use your square tiles to complete the steps for each problem.

Step 1: Construct a rectangle with 3 columns of 4.

Step 2: Separate 2 columns of 4.

Step 3: Write a number bond to show the whole and two parts. Write a repeated addition sentence to match each part of the number bond.

Name ______________________________ Date ______________

Cut out Rectangles A, B, and C. Then, cut according to directions. Answer each of the following using Rectangles A, B, and C.[1]

1. Cut out each row of Rectangle A.

   a. Rectangle A has _____ rows.

   b. Each row has _____ squares.

   c. _____ rows of _____ = _____

   d. Rectangle A has _____ squares.

2. Cut out each column of Rectangle B.

   a. Rectangle B has _____ columns.

   b. Each column has _____ squares.

   c. _____ columns of _____ = _____

   d. Rectangle B has _____ squares.

[1]Note: This Problem Set is used with a template of three identical 2 by 4 arrays. These arrays are labeled as Rectangles A, B, and C.

3. Cut out each square from both Rectangles A and B.

   a. Construct a new rectangle using all 16 squares.

   b. My rectangle has _____ rows of _____.

   c. My rectangle also has _____ columns of _____.

   d. Write two repeated addition number sentences to match your rectangle.

4. Construct a new array using the 24 squares from Rectangles A, B, and C.

   a. My rectangle has _____ rows of _____.

   b. My rectangle also has _____ columns of _____.

   c. Write two repeated addition number sentences to match your rectangle.

Extension: Construct another array using the squares from Rectangles A, B, and C.

   a. My rectangle has _____ rows of _____.

   b. My rectangle also has _____ columns of _____.

   c. Write two repeated addition number sentences to match your rectangle.

Name ______________________________ Date ______________

With your tiles, show 1 rectangle with 12 squares. Complete the sentences below.

I see _____ rows of _____.

In the exact same rectangle, I see _____ columns of _____.

Rectangle A

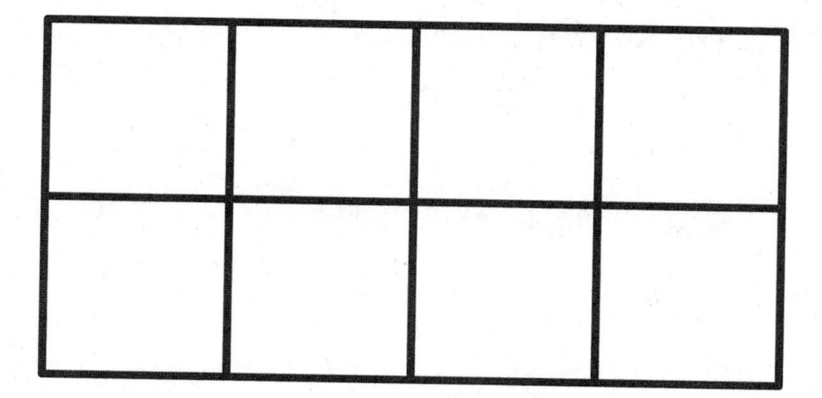

Rectangle B

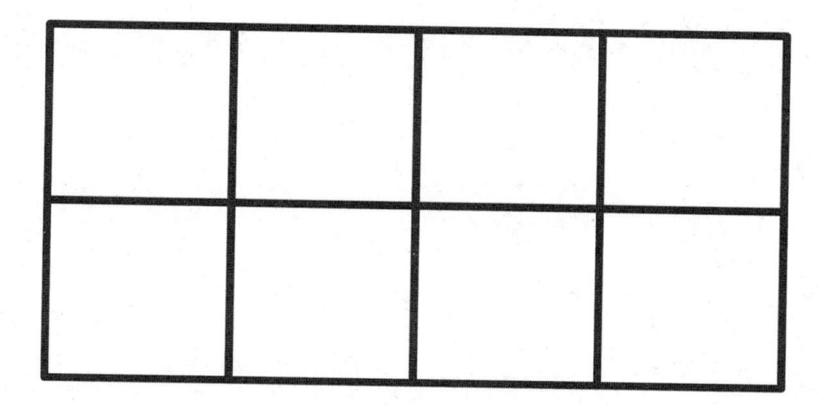

Rectangle C

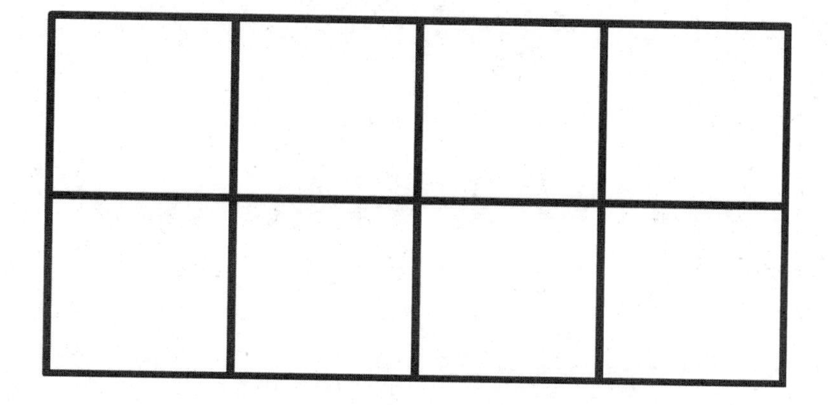

____

rectangles

R (Read the problem carefully.)

Rick is filling his muffin pan with batter. He fills 2 columns of 4. One column of 4 is empty.

a. Draw to show the muffins and the empty column.

b. Write a repeated addition equation to tell how many muffins Rick makes.

W (Write a statement that matches the story.)

____

____

____

Name ______________________ Date ____________

1. Shade in an array with 2 rows of 3.

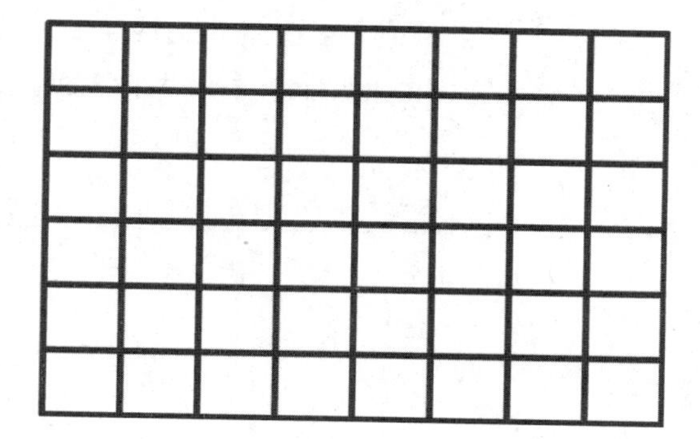

Write a repeated addition equation for the array.

____

2. Shade in an array with 4 rows of 3.

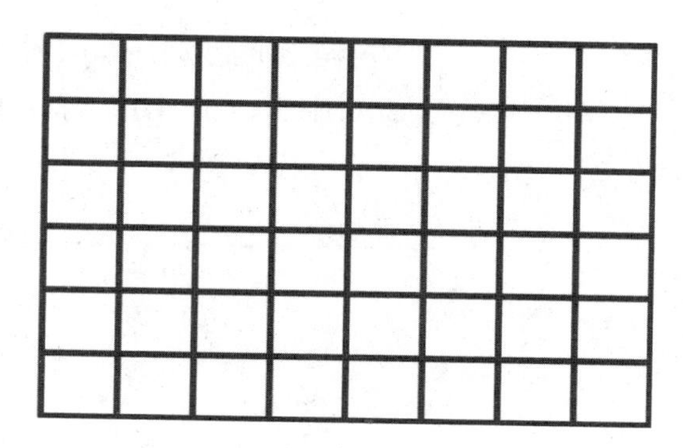

Write a repeated addition equation for the array.

____

3. Shade in an array with 5 columns of 4.

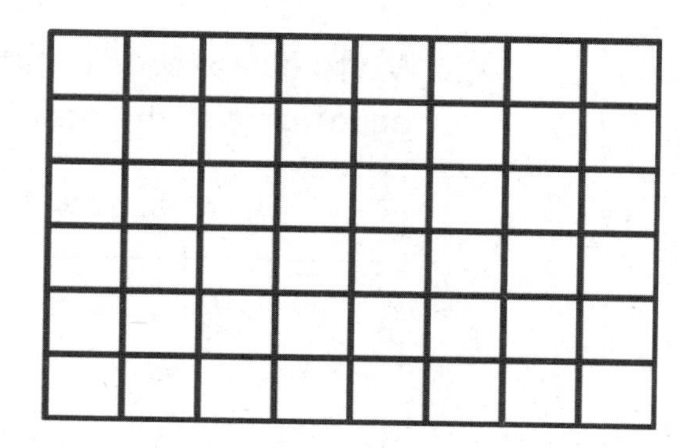

Write a repeated addition equation for the array.

____

4. Draw one more column of 2 to make a new array.

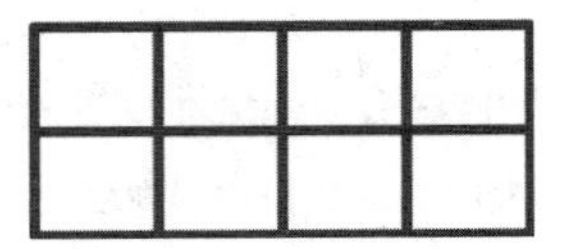

Write a repeated addition equation for the new array.

____

5. Draw one more row of 4 and then one more column to make a new array.

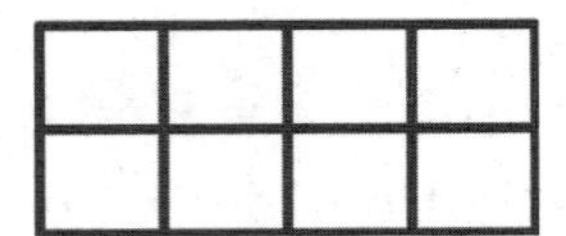

Write a repeated addition equation for the new array.

____

6. Draw one more row and then two more columns to make a new array.

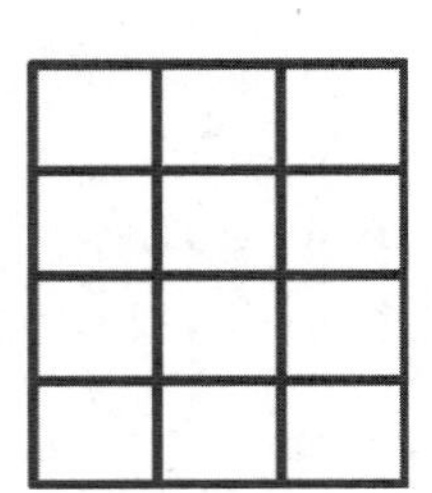

Write a repeated addition equation for the new array.

____

Name ______________________________ Date ______________

Shade in an array with 3 rows of 5.

Write a repeated addition equation for the array.

____

R (Read the problem carefully.)

Rick is baking muffins again. He filled 3 columns of 3 and left one column of 3 empty.

a. Draw a picture to show what the muffin pan looked like. Shade the columns that Rick filled.

b. Write a repeated addition equation to tell how many muffins Rick makes. Then, write a repeated addition equation to tell how many muffins would fit in the whole pan.

W (Write a statement that matches the story.)

____

____

____

Name ____________________________ Date ______________

Use your square tiles and grid paper to complete the following problems.

Problem 1

a. Cut out 10 square tiles.
b. Cut one of your square tiles in half diagonally.
c. Create a design.
d. Shade in your design on grid paper.

Problem 2

a. Use 16 square tiles.
b. Cut two of your square tiles in half diagonally.
c. Create a design.
d. Shade in your design on grid paper.
e. Share your second design with your partner.
f. Check each other's copy to be sure it matches the tile design.

Problem 3

a. Create a 3 by 3 design with your partner in the corner of a new piece of grid paper.
b. With your partner, copy that design to fill the entire paper.

Name ______________________________ Date ______________

Use your square tiles and grid paper to complete the following.

a. Create a design with the paper tiles you used in the lesson.
b. Shade in your design on the grid paper.

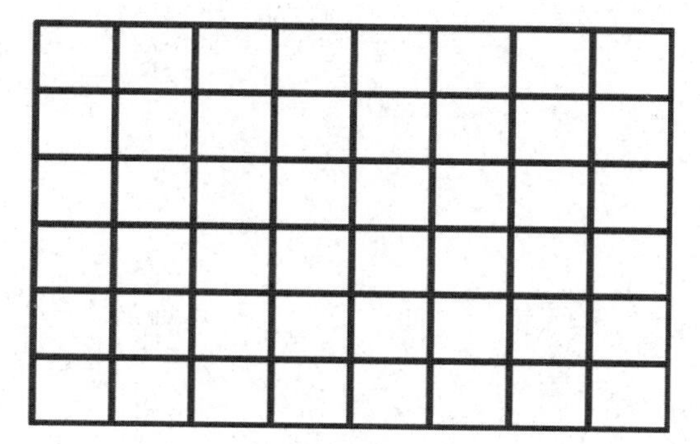

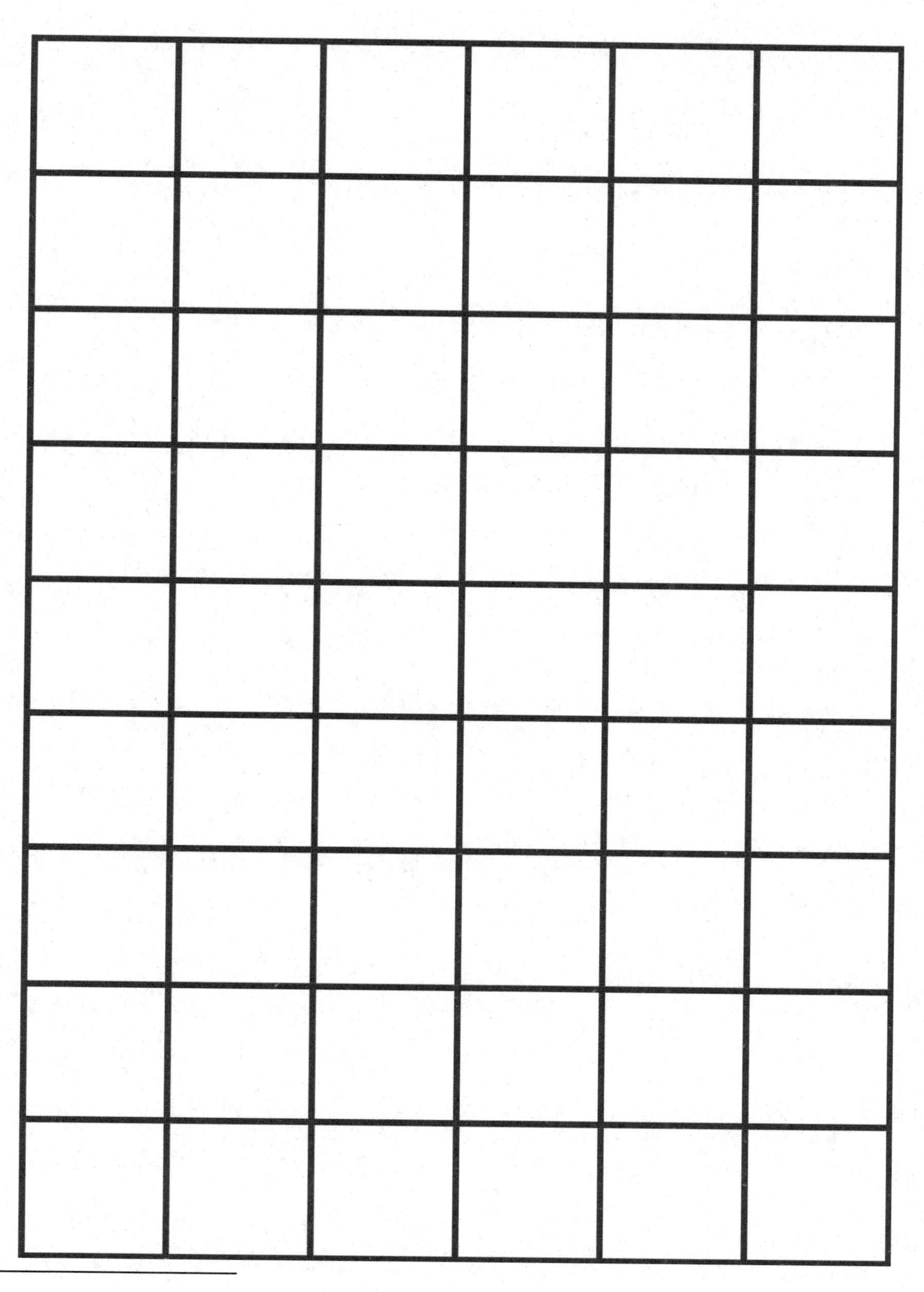

grid paper

Seven students sit on one side of a lunch table. Seven more students sit across from them on the other side of the table.

a. Draw an array to show the students.

b. Write an addition equation that matches the array.

Three more students sit down on each side of the table.

c. Draw an array to show how many students there are now.

d. Write an addition equation that matches the new array.

Name ______________________________ Date ______________

1. Draw to double the group you see. Complete the sentence, and write an addition equation.

a. 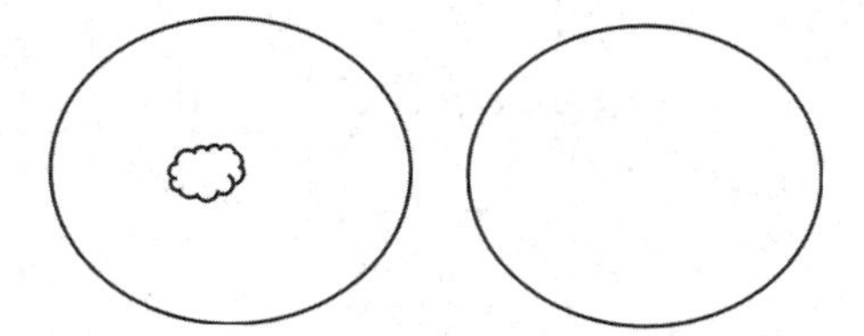

There is _____ cloud in each group.

_______ + _______ = _______

b. 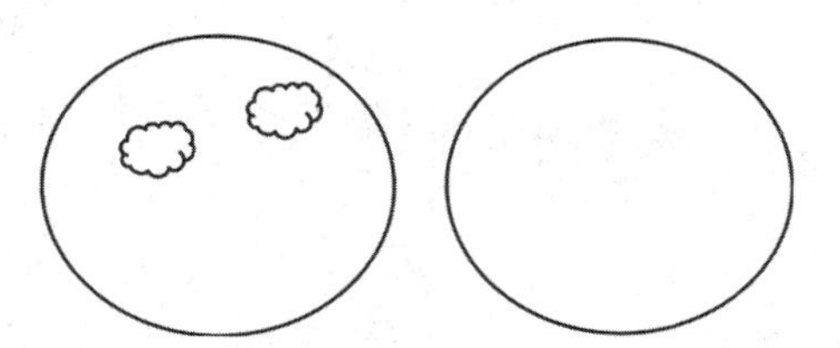

There are _____ clouds in each group.

_______ + _______ = _______

c. 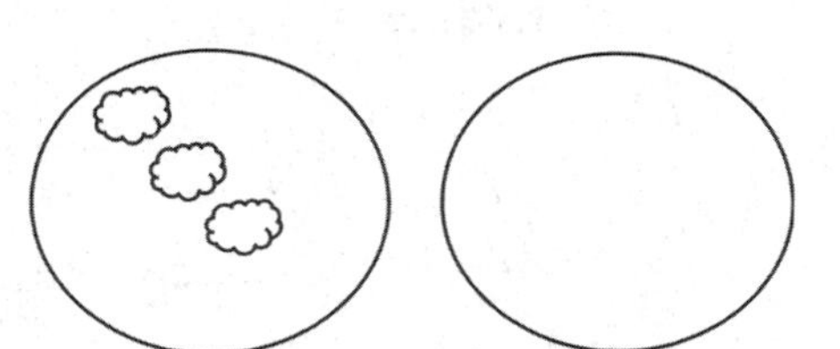

There are _____ clouds in each group.

_______ + _______ = _______

d. 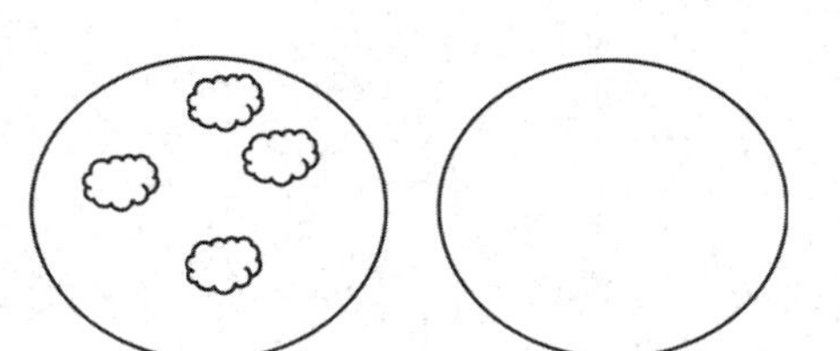

There are _____ clouds in each group.

_______ + _______ = _______

e. 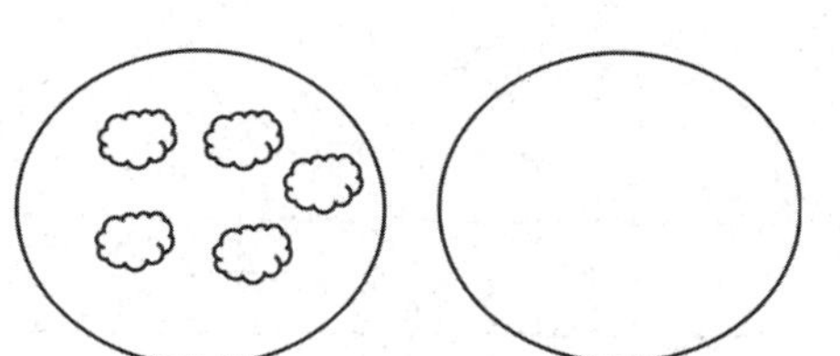

There are _____ clouds in each group.

_______ + _______ = _______

2. Draw an array for each set. Complete the sentences. The first one has been drawn for you.

a. **2 rows of 6**

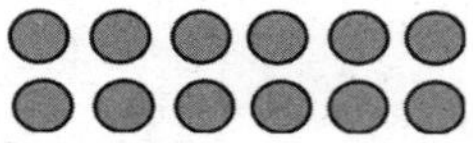

2 rows of 6 = ______

______ + ______ = ______

6 doubled is ______.

b. **2 rows of 7**

2 rows of 7 = ______

______ + ______ = ______

7 doubled is ______.

c. **2 rows of 8**

2 rows of 8 = ______

______ + ______ = ______

8 doubled is ______.

d. **2 rows of 9**

2 rows of 9 = ______

______ + ______ = ______

9 doubled is ______.

e. **2 rows of 10**

2 rows of 10 = ______

______ + ______ = ______

10 doubled is ______.

3. List the totals from Problem 1. ________________________________

List the totals from Problem 2. ________________________________

Are the numbers you have listed even or not even? ______________

Explain in what ways the numbers are the same and different.

Name ______________________________ Date ______________

Draw an array for each set. Complete the sentences.

a. 2 rows of 5

2 rows of 5 = ______

______ + ______ = ______

Circle one: 5 doubled is even/not even.

b. 2 rows of 3

2 rows of 3 = ______

______ + ______ = ______

Circle one: 3 doubled is even/not even.

R (Read the problem carefully.)

Eggs come in cartons of 12. Use pictures, numbers, or words to explain whether 12 is even or not even.

Name ______________________________ Date ______________

1. Pair the objects to decide if the number of objects is even.

Even/Not Even

Even/Not Even

Even/Not Even

2. Draw to continue the pattern of the pairs in the space below until you have drawn 10 pairs.

3. Write the number of dots in each array in Problem 2 in order from least to greatest.

4. Circle the array in Problem 2 that has 2 columns of 7.

5. Box the array in Problem 2 that has 2 columns of 9.

6. Redraw the following sets of dots as columns of two or 2 equal rows.

a. 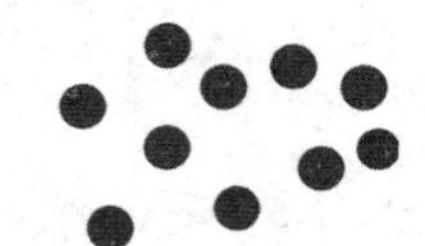

There are __________ dots.

Is _____ an even number? _________

b. 

There are __________ dots.

Is _____ an even number? _________

7. Circle groups of two. Count by twos to see if the number of objects is even.

a. There are _________ twos. There are ________ left over.

b. Count by twos to find the total.

______, ______, ______, ______, ______, ______, ______, ______, ______

c. This group has an even number of objects: True or False

Name ______________________________ Date ______________

Redraw the following sets of dots as columns of two or 2 equal rows.

1.

There are __________ dots.

Is _____ an even number? _________

2.

There are __________ dots.

Is _____ an even number? _________

R (Read the problem carefully.)

Eggs come in cartons of 12. Joanna's mom used 1 egg. Use pictures, numbers, or words to explain whether the amount left is even or odd.

Name _______________ Date _______________

1. Skip-count the columns in the array. The first one has been done for you.

○ ○ ○ ○ ○ ○ ○ ○ ○ ○
○ ○ ○ ○ ○ ○ ○ ○ ○ ○

2 ___ ___ ___ ___ ___ ___ ___ ___ ___

2. a. Solve.

1 + 1 = _______

2 + 2 = _______

3 + 3 = _______

4 + 4 = _______

5 + 5 = _______

6 + 6 = _______

7 + 7 = _______

8 + 8 = _______

9 + 9 = _______

10 + 10 = _______

b. Explain the connection between the array in Problem 1 and the answers in Problem 2(a).

____

____

____

3. a. Fill in the missing numbers on the number path.

20, 22, 24, ____, 28, 30, ____, ____ 36, ____, 40, ____, ____, 46, ____, ____

b. Fill in the odd numbers on the number path.

0, ___, 2, ___, 4, ___, 6, ___, 8, ___, 10, ___, 12, ___, 14, ___, 16, ___, 18, ___, 20, ___

4. Write to identify the **bold** numbers as even or odd. The first one has been done for you.

a.	b.	c.
6 + 1 = **7** even + 1 = odd	**24** + 1 = **25** ____ + 1 = ____	**30** + 1 = **31** ____ + 1 = ____
d. **6** – 1 = **5** ____ – 1 = ____	e. **24** – 1 = **23** ____ – 1 = ____	f. **30** – 1 = **29** ____ – 1 = ____

5. Are the **bold** numbers even or odd? Circle the answer, and explain how you know.

a. **28** even/odd	Explanation:
b. **39** even/odd	Explanation:
c. **45** even/odd	Explanation:
d. **50** even/odd	Explanation:

Name ______________________________ Date ________________

Are the **bold** numbers even or odd? Circle the answer, and explain how you know.

a. **18** even/odd	Explanation:
b. **23** even/odd	Explanation:

R (Read the problem carefully.)

Mrs. Boxer has 11 boys and 9 girls at a Grade 2 party.

a. Write the equation to show the total number of people.

b. Are the addends even or odd?

c. Mrs. Boxer wants to pair everyone up for a game. Does she have the right number of people for everyone to have a partner?

D (Draw a picture.)

W (Write and solve an equation.)

W (Write a statement that matches the story.)

____

____

____

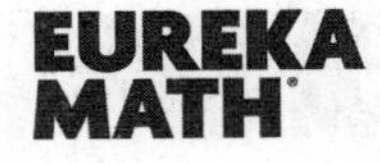

Name ______________________________ Date ______________

1. Use the objects to create an array.

a. 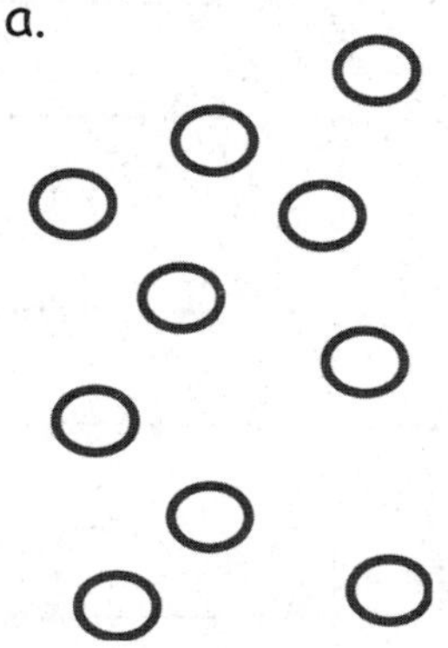	Array There are an even/odd (circle one) number of circles.	Redraw your picture with 1 *less* circle. There are an even/odd (circle one) number of circles.
b. 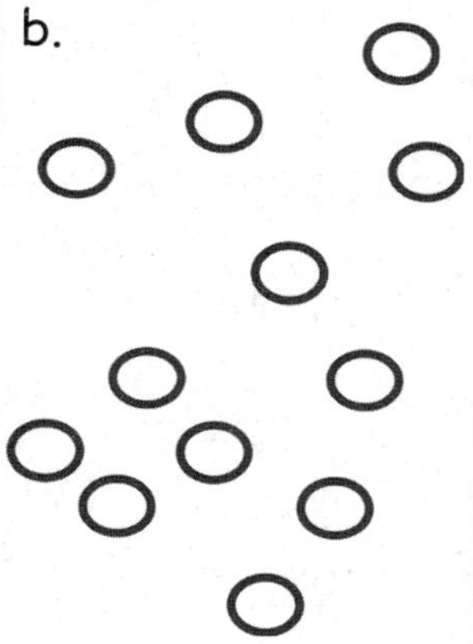	Array There are an even/odd (circle one) number of circles.	Redraw your picture with 1 *more* circle. There are an even/odd (circle one) number of circles.
c. 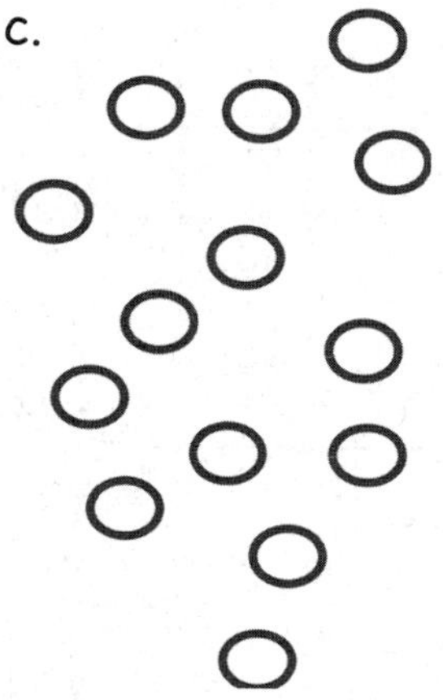	Array There are an even/odd (circle one) number of circles.	Redraw your picture with 1 *less* circle. There are an even/odd (circle one) number of circles.

2. Solve. Tell if each number is odd (O) or even (E). The first one has been done for you.

a. 6 + 4 = 10

E + E = E

b. 17 + 2 = ______

______ + ______ = ______

c. 11 + 13 = ______

______ + ______ = ______

d. 14 + 8 = ______

______ + ______ = ______

e. 3 + 9 = ______

______ + ______ = ______

f. 5 + 14 = ______

______ + ______ = ______

3. Write two examples for each case. Write if your answers are even or odd. The first one has been started for you.

a. Add an even number to an even number.

32 + 8 = 40 even ______________________

b. Add an odd number to an even number.

______________________ ______________________

c. Add an odd number to an odd number.

______________________ ______________________

Name ______________________________ Date ______________

Use the objects to create an array.

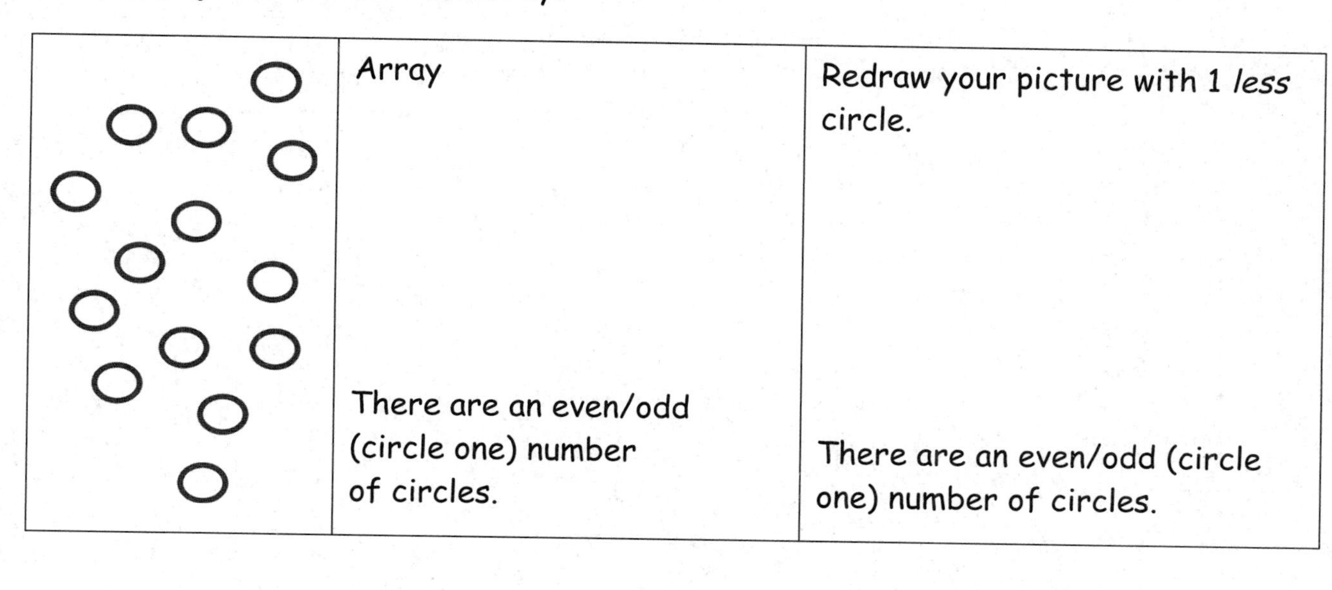

	Array	Redraw your picture with 1 *less* circle.
	There are an even/odd (circle one) number of circles.	There are an even/odd (circle one) number of circles.

Grade 2
Module 7

R (Read the problem carefully.)

There are 24 penguins sliding on the ice. There are 18 whales splashing in the ocean. How many more penguins than whales are there?

D (Draw a picture.)

W (Write and solve an equation.)

W (Write a statement that matches the story.)

___

___

___

Name ______________________________ Date ________________

1. Count and categorize each picture to complete the table with tally marks.

No Legs	2 Legs	4 Legs

2. Count and categorize each picture to complete the table with numbers.

Fur	Feathers

3. Use the Animal Habitats table to answer the following questions.

Animal Habitats		
Forest	**Wetlands**	**Grasslands**
~~||||~~ |	~~||||~~	~~||||~~ ~~||||~~ ||||

a. How many animals have habitats on grasslands and wetlands? ______

b. How many fewer animals have forest habitats than grasslands habitats? ______

c. How many more animals would need to be in the forest category to have the same number as animals in the grasslands category? ______

d. How many total animal habitats were used to create this table? ______

4. Use the Animal Classification table to answer the following questions about the types of animals Ms. Lee's second-grade class found in the local zoo.

Animal Classification			
Birds	**Fish**	**Mammals**	**Reptiles**
6	5	11	3

a. How many animals are birds, fish, or reptiles? ______

b. How many more birds and mammals are there than fish and reptiles? ______

c. How many animals were classified? ______

d. How many more animals would need to be added to the chart to have 35 animals classified? ______

e. If 5 more birds and 2 more reptiles were added to the table, how many fewer reptiles would there be than birds? ______

Name _______________________________ Date _______________

Use the Animal Classification table to answer the following questions about the types of animals at the local zoo.

Animal Classification			
Birds	**Fish**	**Mammals**	**Reptiles**
9	4	17	8

1. How many animals are birds, fish, or reptiles? ______

2. How many more mammals are there than fish? ______

3. How many animals were classified? ______

4. How many more animals would need to be added to the chart to have 45 animals classified? ______

R (Read the problem carefully.)

Gemma is counting animals in the park. She counts 16 robins, 19 ducks, and 17 squirrels. How many more robins and ducks did Gemma count than squirrels?

D (Draw a picture.)

W (Write and solve an equation.)

W (Write a statement that matches the story.)

____

____

____

Name ______________________________ Date ______________

1. Use grid paper to create a picture graph below using data provided in the table. Then, answer the questions.

Central Park Zoo Animal Classification			
Birds	**Fish**	**Mammals**	**Reptiles**
6	5	11	3

Title: ______________

Legend: ______________

a. How many more animals are mammals than fish? ______

b. How many more animals are mammals and fish than birds and reptiles? ______

c. How many fewer animals are reptiles than mammals? ______

d. Write and answer your own comparison question based on the data.

Question: ______________________________

Answer: ______________________________

2. Use the table below to create a picture graph in the space provided.

Animal Habitats		
Desert	**Tundra**	**Grasslands**
卌 |	卌	卌 卌 ||||

Title: ______________________

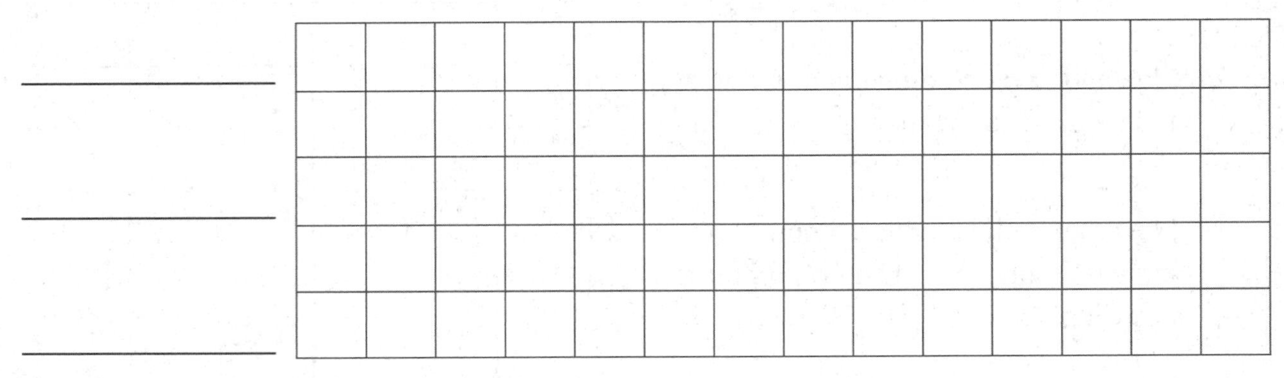

Legend: ______________________

a. How many more animal habitats are in the grasslands than in the desert? ______

b. How many fewer animal habitats are in the tundra than in the grasslands and desert combined? ______

c. Write and answer your own comparison question based on the data.

Question: ____

Answer: ____

Name ______________________________ Date ______________

Use grid paper to create a picture graph below using data provided in the table. Then, answer the questions.

Fairview Park Zoo Animal Classification			
Birds	Fish	Mammals	Reptiles
8	4	12	5

a. How many more animals are mammals than birds? ________

b. How many more animals are mammals and reptiles than birds and fish? ________

c. How many fewer animals are fish than birds? ________

Title: ______________________

____ ____ ____ ____

Legend: ______________________

Legend:

Legend:

vertical and horizontal picture graphs

Legend: ____________________

vertical picture graph

a. Use the tally chart to fill in the picture graph.

b. Draw a tape diagram to show how many more books Jose read than Laura.

c. If Jose, Laura, and Linda read 21 books altogether, how many books did Linda read?

Number of Books Read

Jose	Laura	Linda
~~IIII~~ III	~~IIII~~	

d. Complete the tally chart and the graph.

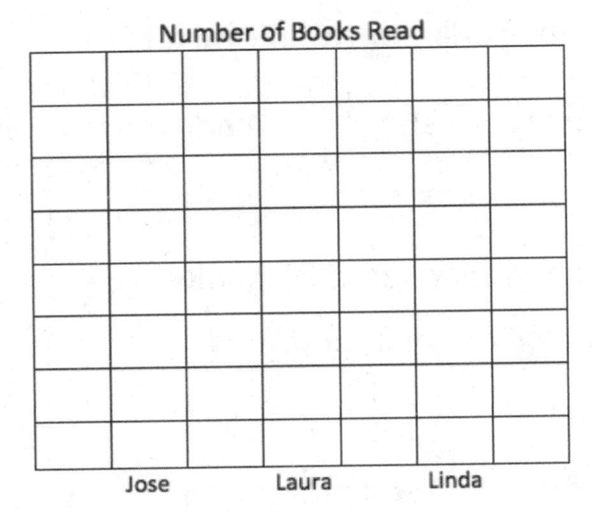

Each ● stands for 1 book.

Name ____________________ Date __________

1. Complete the bar graph below using data provided in the table. Then, answer the questions about the data.

Animal Classification			
Birds	Fish	Mammals	Reptiles
6	5	11	3

Title: ____________________

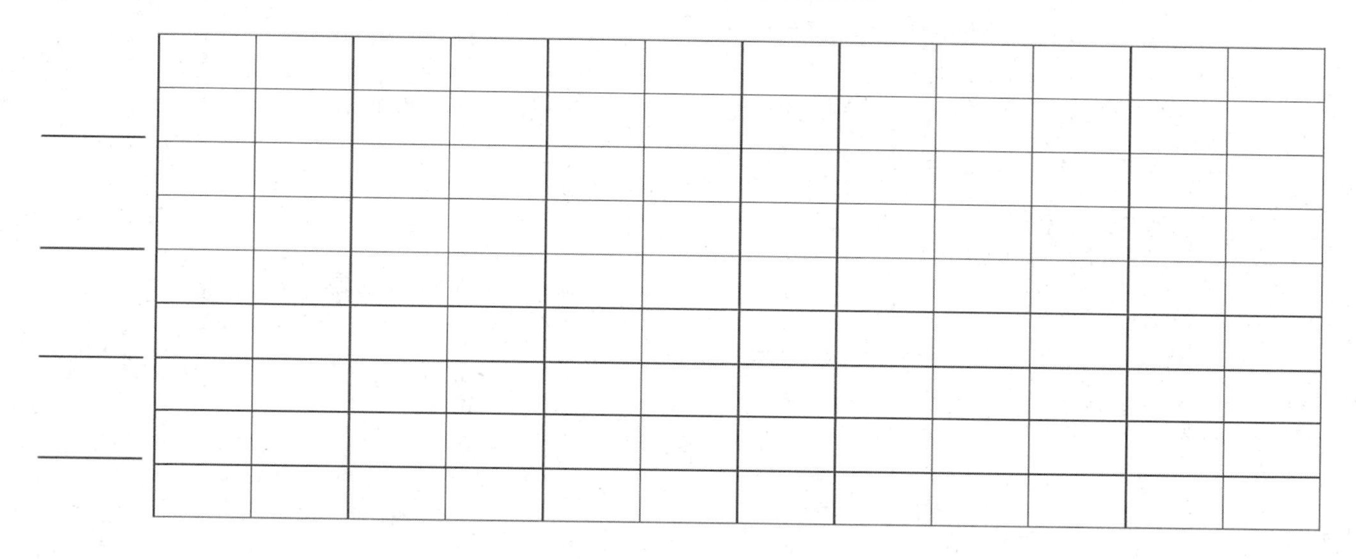

0 __ __ __ __ __ __ __ __ __ __ __ __

a. How many more animals are birds than reptiles? ______

b. How many more birds and mammals are there than fish and reptiles? ______

c. How many fewer animals are reptiles and fish than mammals? ______

d. Write and answer your own comparison question based on the data.

Question: ____________________

Answer: ____________________

Animal Habitats		
Desert	**Arctic**	**Grasslands**
𝍸 |	𝍸	𝍸 𝍸 ||||

2. Complete the bar graph below using data provided in the table.

Title: ______________________

14
13
12
11
10
9
8
7
6
5
4
3
2
1
0

_____ _____ _____

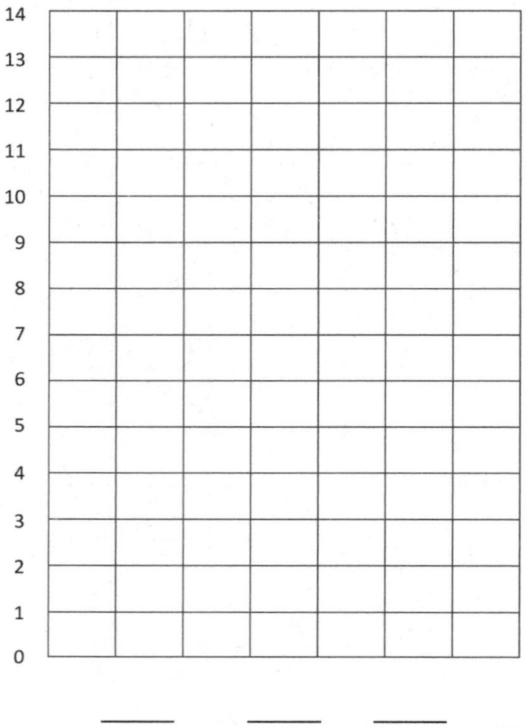

a. How many more animals live in the grasslands and arctic habitats combined than in the desert? _______

b. If 3 more grasslands animals and 4 more arctic animals are added to the graph, how many grasslands and arctic animals would there be? _______

c. If 3 animals were removed from each category, how many animals would there be? _______

d. Write your own comparison question based on the data, and answer it.

Question: ____

Answer: ____

Name ______________________________ Date ______________

Complete the bar graph below using data provided in the table. Then, answer the questions about the data.

Animal Classification			
Birds	Fish	Mammals	Reptiles
7	12	8	6

Title: ______________________________

0 __ __ __ __ __ __ __ __ __ __ __ __ __ __

a. How many more animals are fish than reptiles? ______

b. How many more fish and mammals are there than birds and reptiles? ______

0

Title:

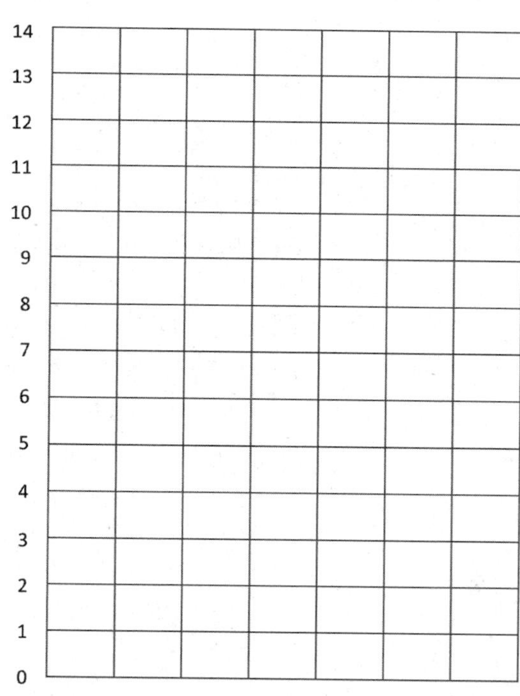

horizontal and vertical bar graphs

After a trip to the zoo, Ms. Anderson's students voted on their favorite animals. Use the bar graph to answer the following questions.

a. Which animal got the fewest votes?

b. Which animal got the most votes?

c. How many more students liked Komodo dragons than koala bears?

d. Later, two students changed their votes from koala bear to snow leopard. What was the difference between koala bears and snow leopards then?

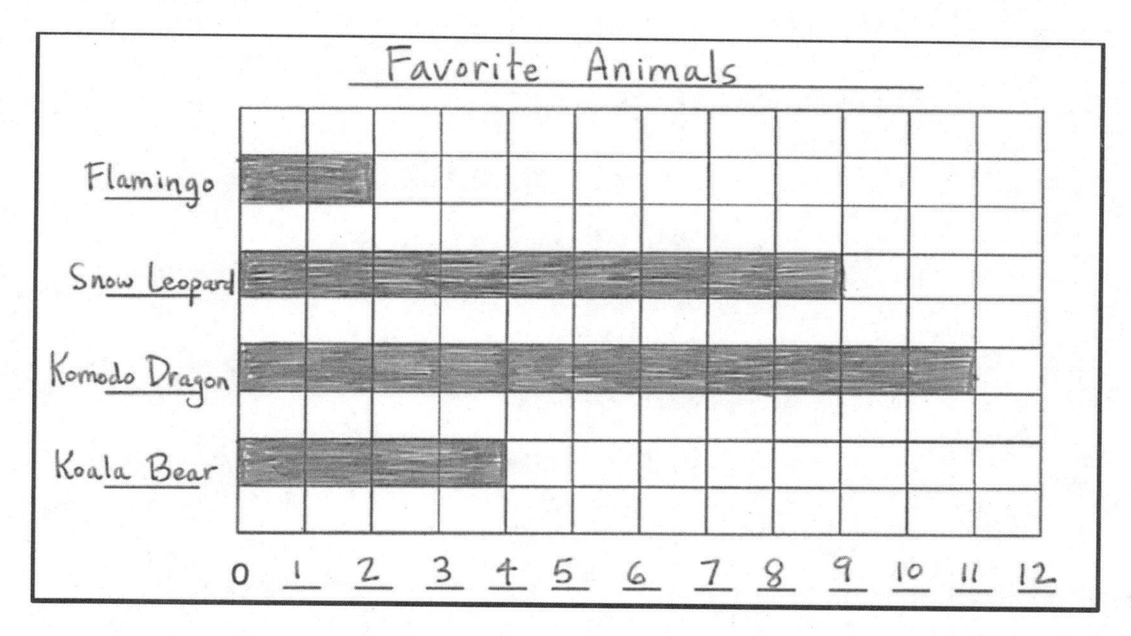

a.

b.

c.

d.

Name ______________________________ Date ______________

1. Complete the bar graph using the table with the types of bugs Alicia counted in the park. Then, answer the following questions.

Types of Bugs			
Butterflies	Spiders	Bees	Grasshoppers
5	14	12	7

Title: ______________________

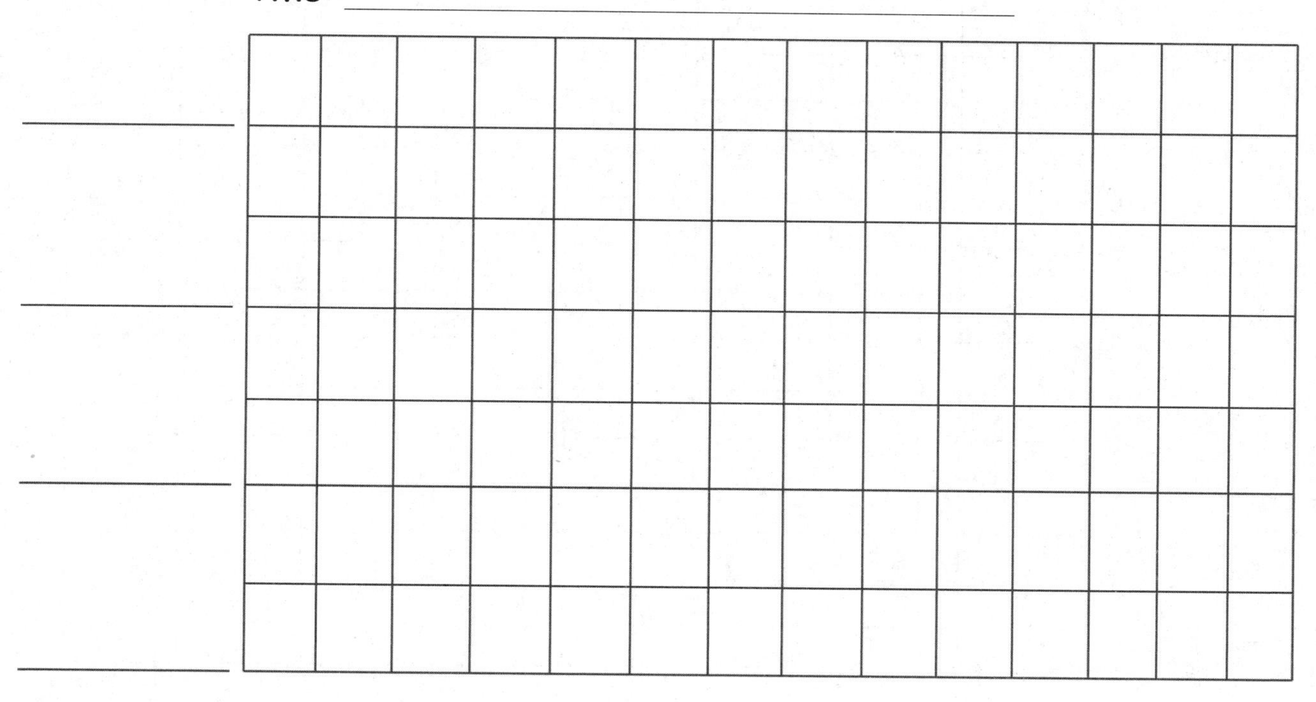

0 __ __ __ __ __ __ __ __ __ __ __ __ __ __

a. How many butterflies were counted in the park? ______

b. How many more bees than grasshoppers were counted in the park? ______

c. Which bug was counted twice as many times as grasshoppers? ______

d. How many bugs did Alicia count in the park? ______

e. How many fewer butterflies than bees and grasshoppers were counted in the park? ______

2. Complete the bar graph with labels and numbers using the number of farm animals on O'Brien's farm.

O'Brien's Farm Animals			
Goats	Pigs	Cows	Chickens
13	15	7	8

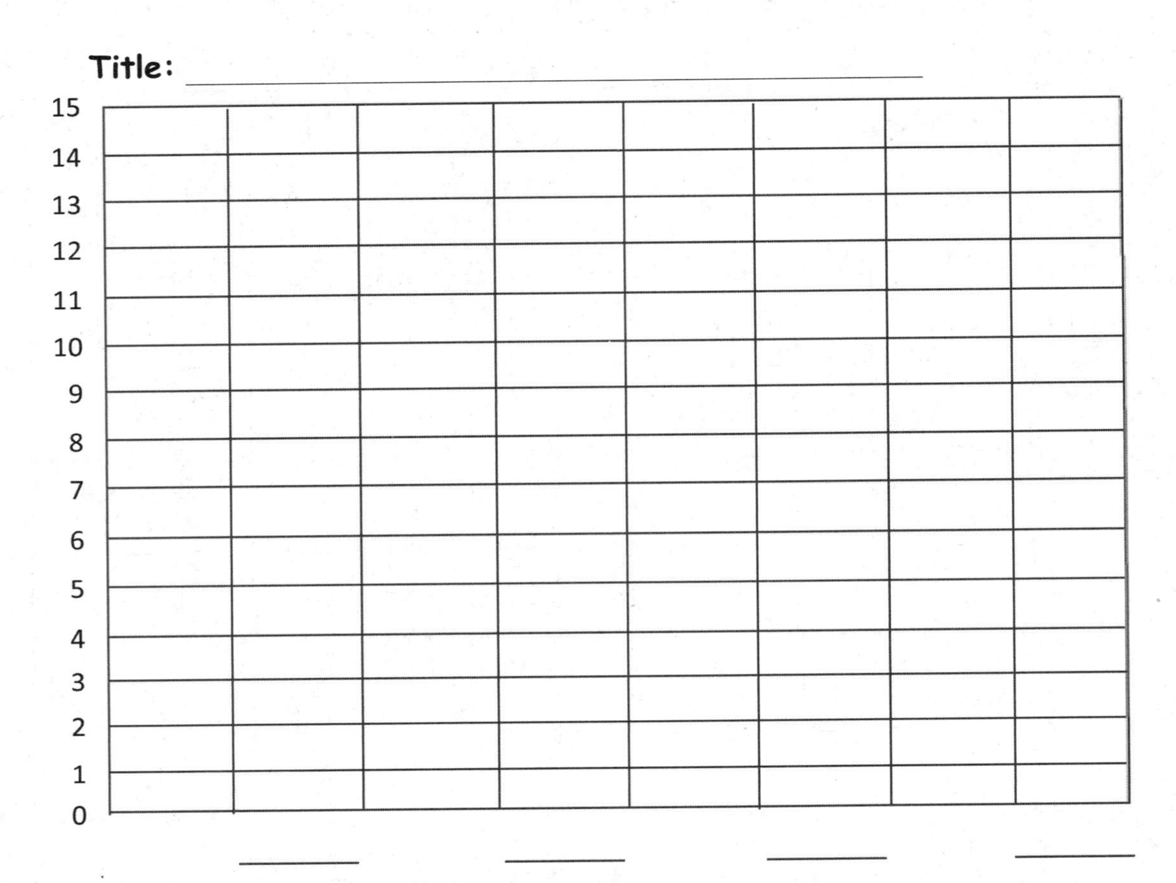

a. How many more pigs than chickens are on O'Brien's farm? ________

b. How many fewer cows than goats are on O'Brien's farm? ________

c. How many fewer chickens than goats and cows are on O'Brien's farm? ________

d. Write a comparison question that can be answered using the data on the bar graph.

____

Name ______________________________ Date ______________

Complete the bar graph using the table with the types of bugs Jeremy counted in his backyard. Then, answer the following questions.

Types of Bugs			
Butterflies	Spiders	Bees	Grasshoppers
4	8	10	9

Title: ______________________________

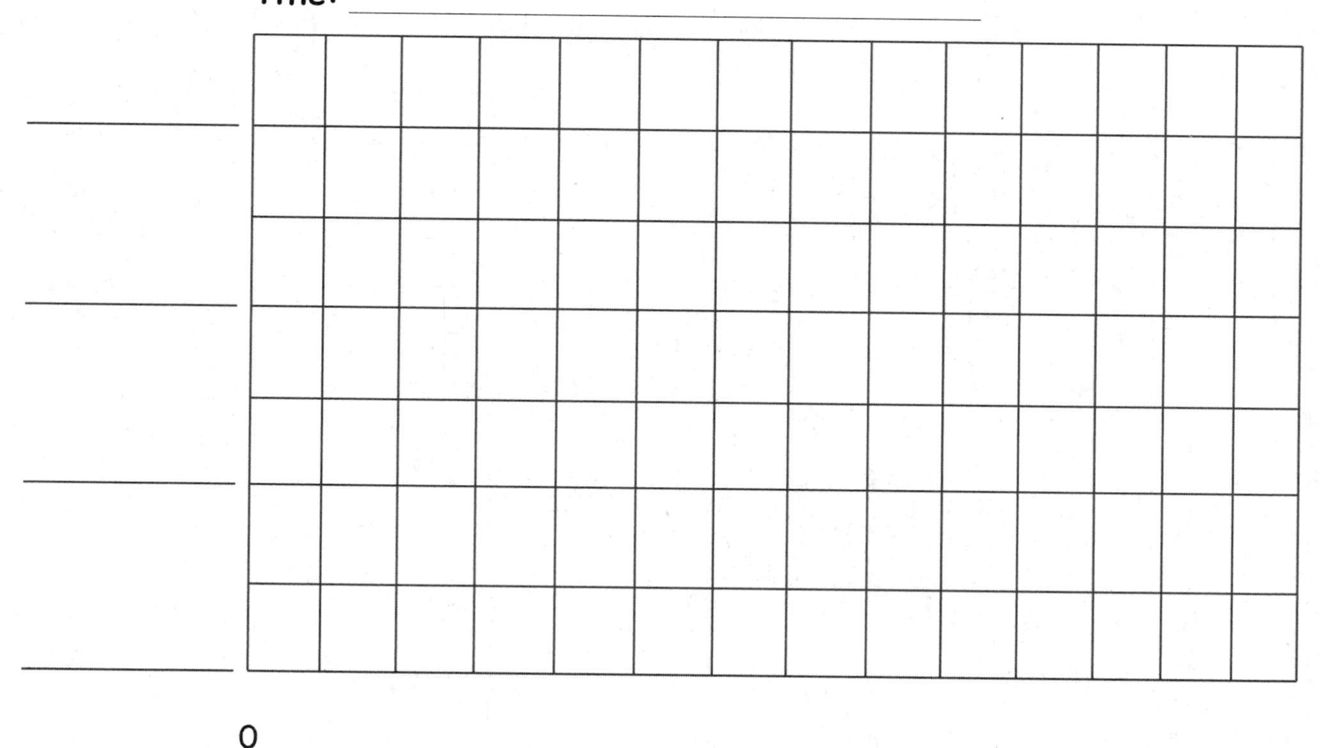

a. How many more spiders and grasshoppers were counted than bees and butterflies?

____

b. If 5 more butterflies were counted, how many bugs would have been counted?

____

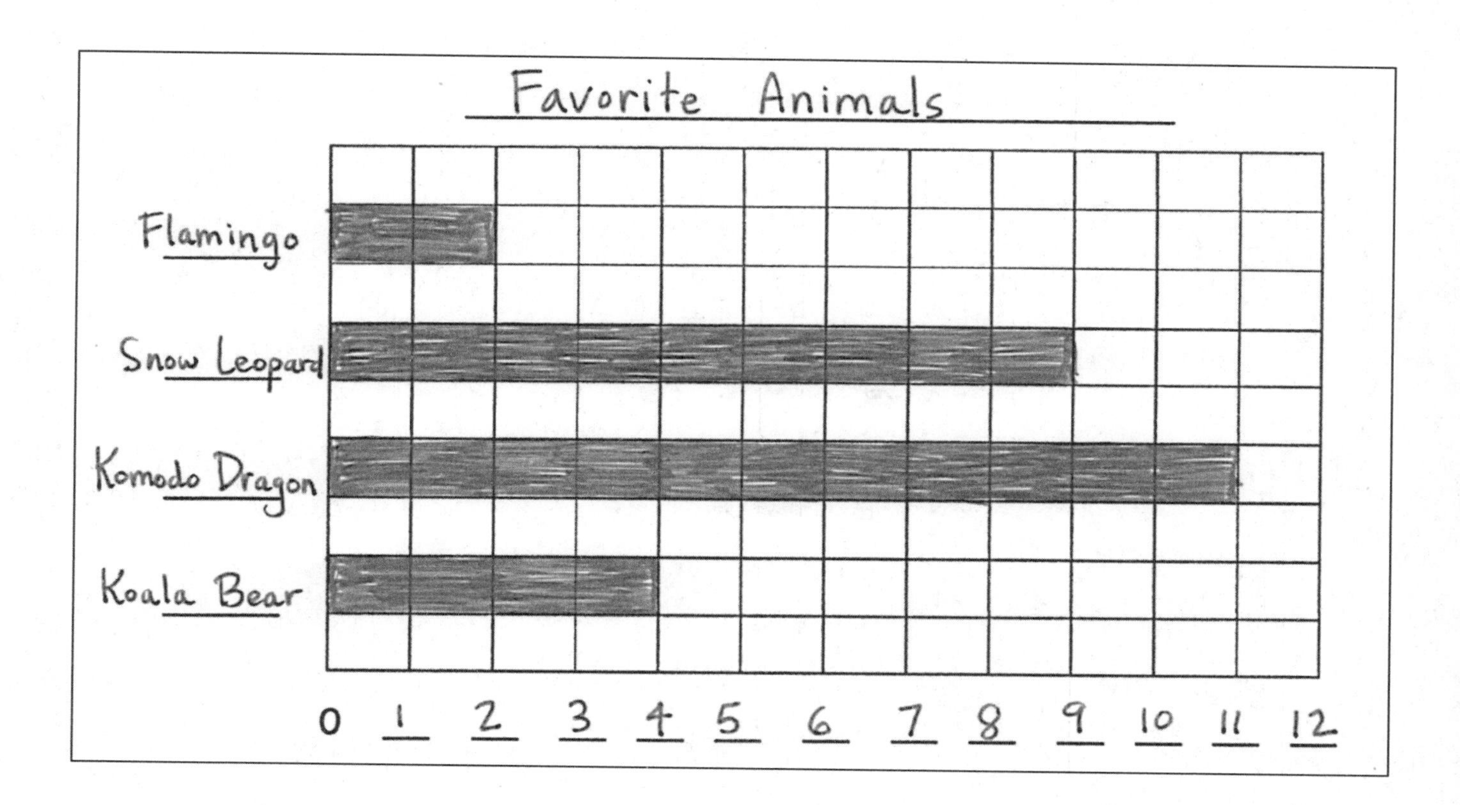

favorite animals bar graph

R (Read the problem carefully.)

Rita has 19 more pennies than Carlos. Rita has 27 pennies. How many pennies does Carlos have?

D (Draw a picture.)

W (Write and solve an equation.)

W (Write a statement that matches the story.) no

Name ______________________________ Date ______________

Callista saved pennies. Use the table to complete the bar graph. Then, answer the following questions.

Pennies Saved			
Saturday	Sunday	Monday	Tuesday
15	10	4	7

Title: ______________________

15
14
13
12
11
10
9
8
7
6
5
4
3
2
1
0

_______ _______ _______ _______

a. How many pennies did Callista save in all? _______

b. Her sister saved 18 fewer pennies. How many pennies did her sister save? _______

c. How much more money did Callista save on Saturday than on Monday and Tuesday? _______

d. How will the data change if Callista doubles the amount of money she saved on Sunday? ______________________________

e. Write a comparison question that can be answered using the data on the bar graph.

____

Name ______________________________ Date ______________

A group of friends counted their nickels. Use the table to complete the bar graph. Then, answer the following questions.

Amount of Nickels			
Annie	Scarlett	Remy	LaShay
5	11	8	14

Title: ______________________________

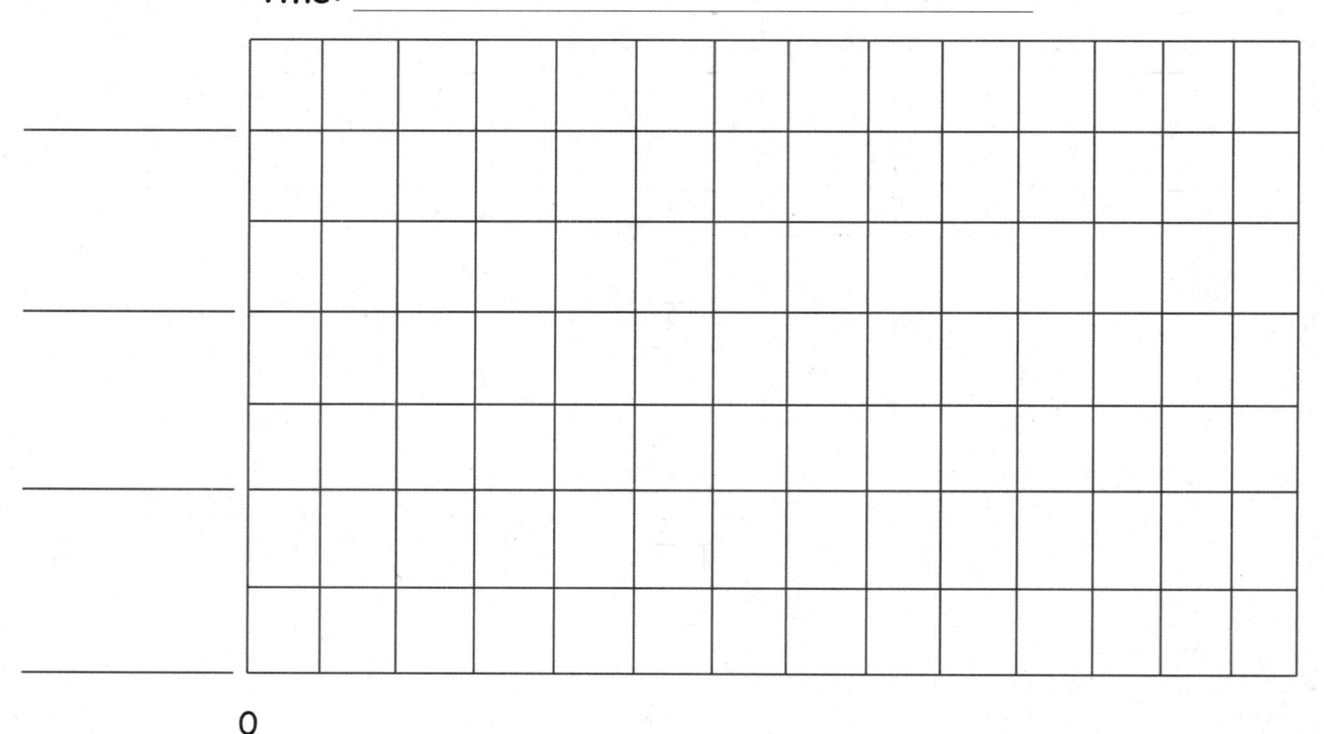

a. How many nickels do the children have in all? _____

b. What is the total value of Annie's and Remy's coins? _____

c. How many fewer nickels does Remy have than LaShay? _____

d. Who has less money, Annie and Scarlett or Remy and LaShay? ____________

e. Write a comparison question that can be answered using the data on the bar graph.

____

Name ______________________________ Date ______________

1. Design a survey, and collect the data.
2. Label and fill in the table.
3. Use the table to label and complete the bar graph.
4. Write questions based on the graph, and then let students use your graphs to answer them.

    a. ____

    b. ____

    c. ____

    d. ____

Name ______________________________ Date ______________

1. Use the table to complete the bar graph. Then, answer the following questions.

Number of Dimes

Emily	Andrew	Thomas	Ava
8	12	6	13

Title: ______________________________

a. How many more dimes does Andrew have than Emily? ______

b. How many fewer dimes does Thomas have than Ava and Emily? ______

c. Circle the pair with more dimes, Emily and Ava or Andrew and Thomas.
How many more? ______

d. What is the total number of dimes if all the students combine all their money?

____

2. Use the table to complete the bar graph. Then, answer the following questions.

Number of Dimes Donated

Madison	Robin	Benjamin	Miguel
12	10	15	13

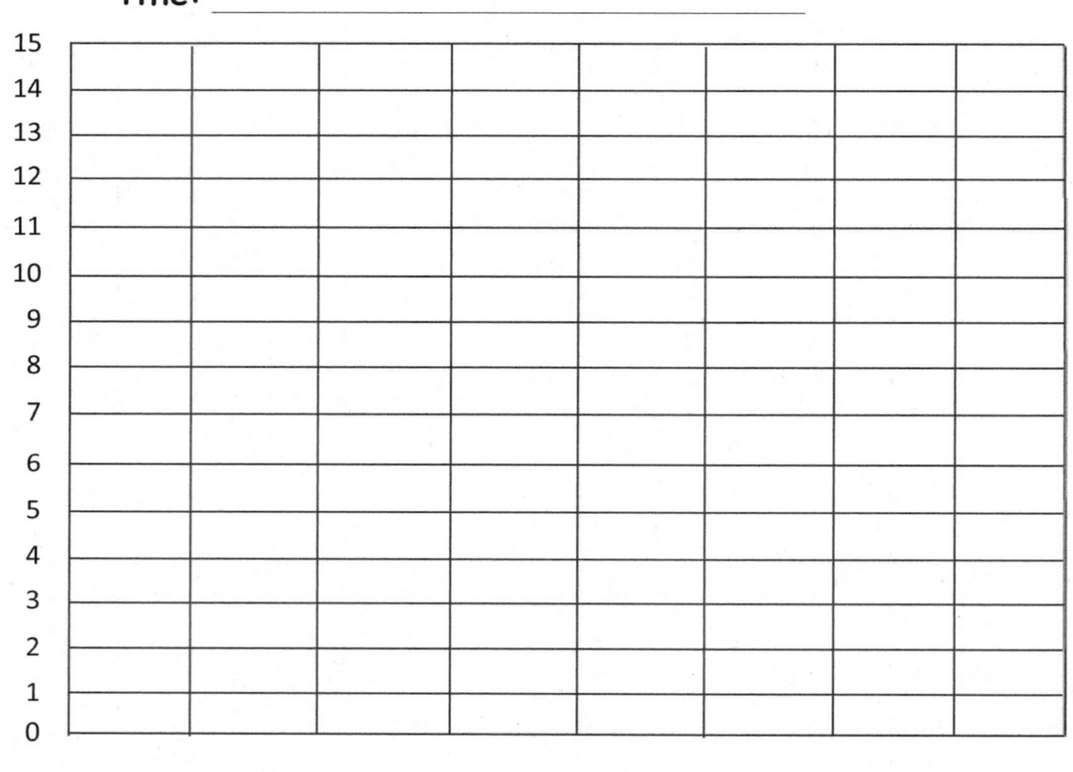

a. How many more dimes did Miguel donate than Robin? ________

b. How many fewer dimes did Madison donate than Robin and Benjamin? ________

c. How many more dimes are needed for Miguel to donate the same as Benjamin and Madison? ________

d. How many dimes were donated? ________

Name ______________________________ Date ______________

Use the table to complete the bar graph. Then, answer the following questions.

Number of Dimes

Lacy	Sam	Stefanie	Amber
6	11	9	14

Title: ______________________________

a. How many more dimes does Amber have than Stefanie? ______

b. How many dimes will Sam and Lacy need to save to equal Stefanie and Amber?

______

R (Read the problem carefully.)

Sarah is saving money in her piggy bank. So far, she has 3 dimes, 1 quarter, and 8 pennies.

a. How much money does Sarah have?

b. How much more does she need to have a dollar?

D (Draw a picture.)

W (Write and solve an equation.)

W (Write a statement that matches the story.)

a. ______________________________

____

b. ______________________________

____

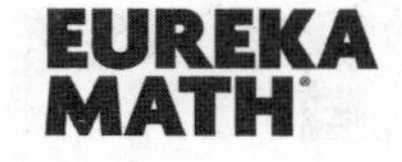

Name ________________________________ Date ______________

Count or add to find the total value of each group of coins.
Write the value using the ¢ or $ symbol.

1. 	________
2. 	________
3. 	________
4. 	________
5. 	________
6. 	________
7. 	________

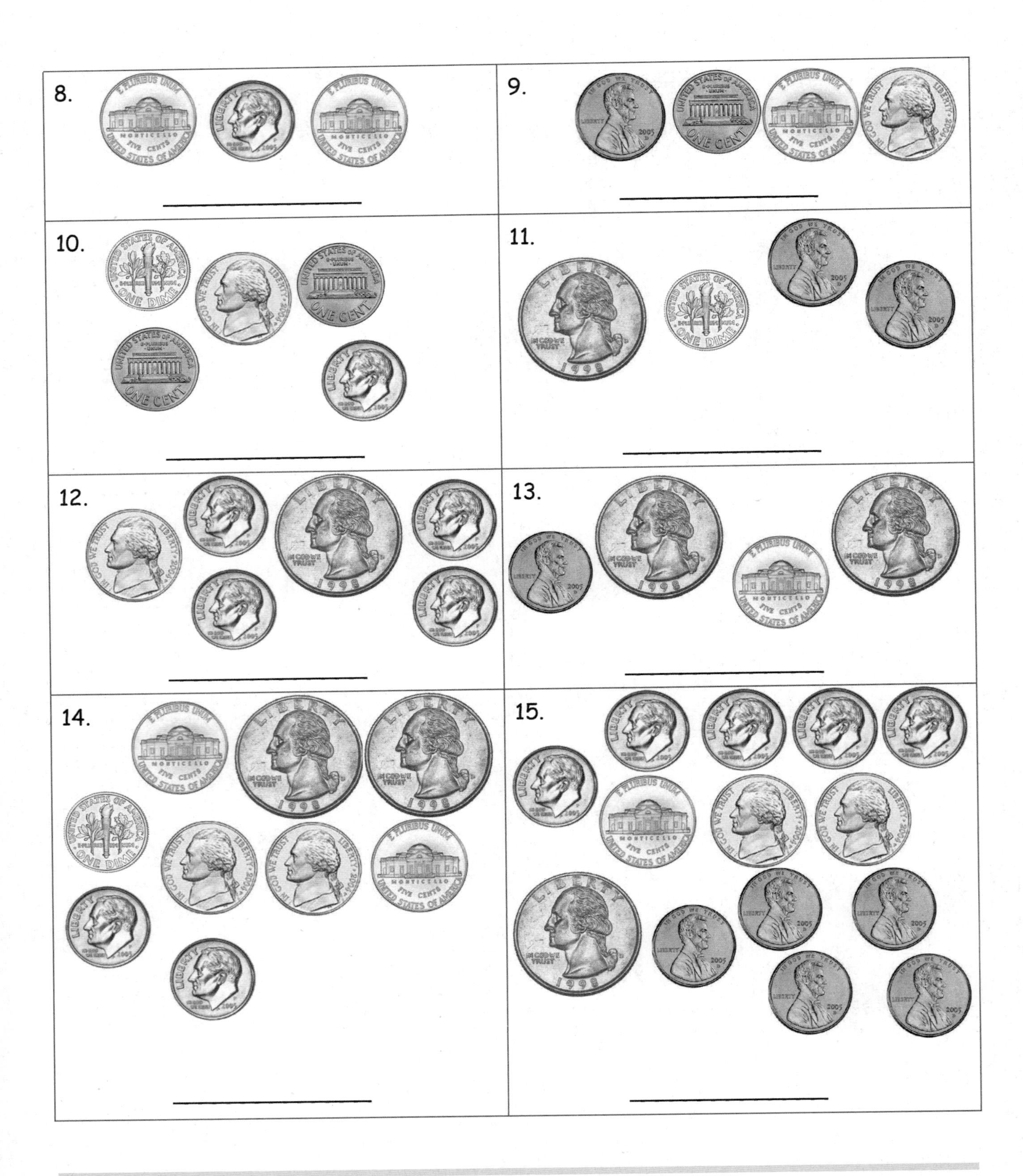
8.
9.
10.
11.
12.
13.
14.
15.

Name ______________________________ Date ________________

Count or add to find the total value of each group of coins.

Write the value using the ¢ or $ symbol.

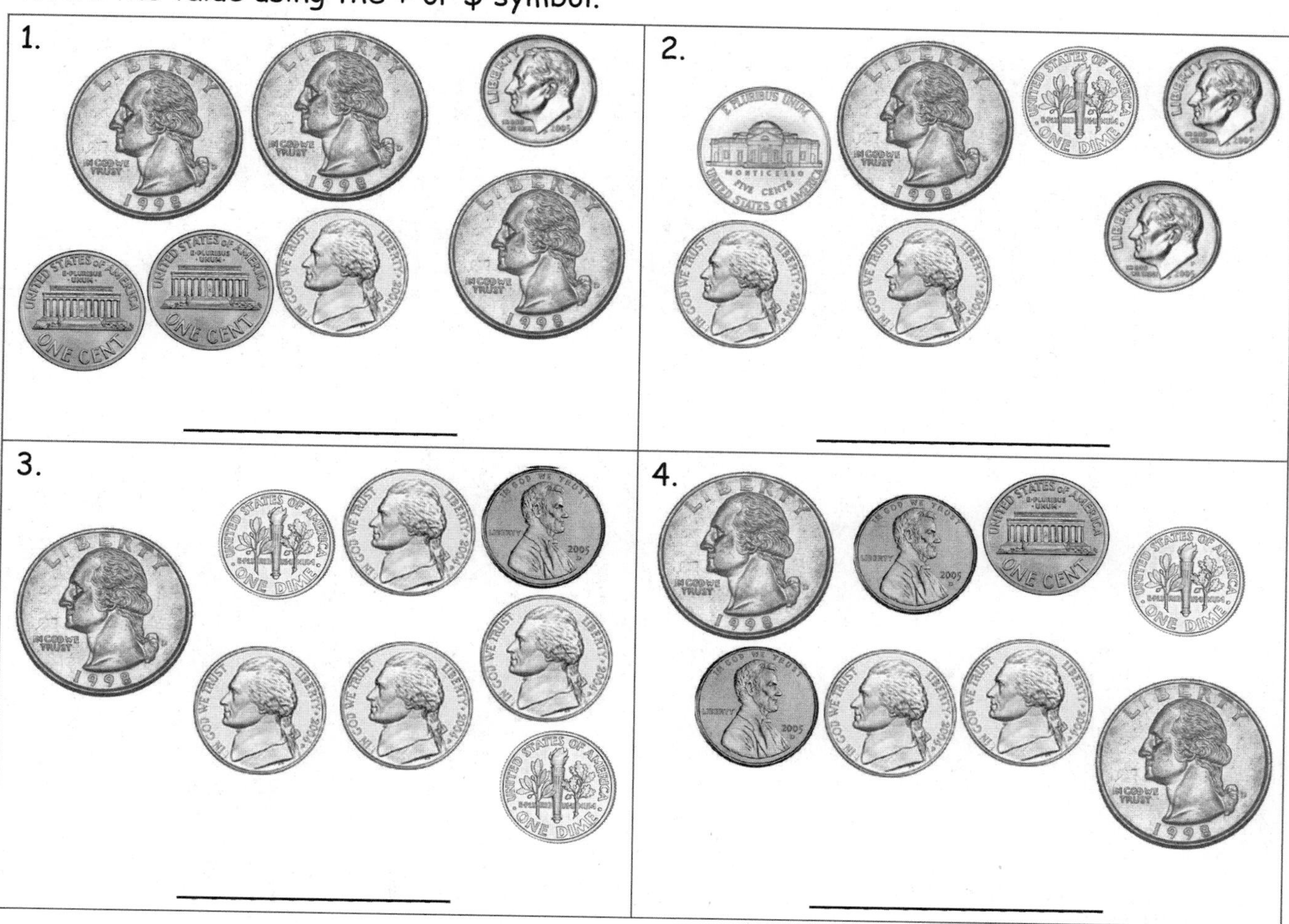

R (Read the problem carefully.)

Danny has 2 dimes, 1 quarter, 3 nickels, and 5 pennies.

a. What is the total value of Danny's coins?

b. Show two different ways that Danny might add to find the total.

D (Draw a picture.)

W (Write and solve an equation.)

W (Write a statement that matches the story.)

a. ______________________________

____

b. ______________________________

____

Name ____________________ Date __________

Solve.

1. Grace has 3 dimes, 2 nickels, and 12 pennies. How much money does she have?

2. Lisa has 2 dimes and 4 pennies in one pocket and 4 nickels and 1 quarter in the other pocket. How much money does she have in all?

3. Mamadou found 39 cents in the sofa last week. This week, he found 2 nickels, 4 dimes, and 5 pennies. How much money does Mamadou have altogether?

4. Emanuel had 53 cents. He gave 1 dime and 1 nickel to his brother. How much money does Emanuel have left?

5. There are 2 quarters and 14 pennies in the top drawer of the desk and 7 pennies, 2 nickels, and 1 dime in the bottom drawer. What is the total value of the money in both drawers?

6. Ricardo has 3 quarters, 1 dime, 1 nickel, and 4 pennies. He gave 68 cents to his friend. How much money does Ricardo have left?

Name ____________________ Date ____________

Solve.

1. Greg had 1 quarter, 1 dime, and 3 nickels in his pocket. He found 3 nickels on the sidewalk. How much money does Greg have?

2. Robert gave Sandra 1 quarter, 5 nickels, and 2 pennies. Sandra already had 3 pennies and 2 dimes. How much money does Sandra have now?

R (Read the problem carefully.)

Kiko's brother says that he will trade her 2 quarters, 4 dimes, and 2 nickels for a one-dollar bill. Is this a fair trade? How do you know?

D (Draw a picture.)

W (Write and solve an equation.)

W (Write a statement that matches the story.)

Name ______________________________ Date ______________

Solve.

1. Patrick has 1 ten-dollar bill, 2 five-dollar bills, and 4 one-dollar bills. How much money does he have?

2. Susan has 2 five-dollar bills and 3 ten-dollar bills in her purse and 11 one-dollar bills in her pocket. How much money does she have in all?

3. Raja has $60. He gave 1 twenty-dollar bill and 3 five-dollar bills to his cousin. How much money does Raja have left?

4. Michael has 4 ten-dollar bills and 7 five-dollar bills. He has 3 more ten-dollar bills and 2 more five-dollar bills than Tamara. How much money does Tamara have?

5. Antonio had 4 ten-dollar bills, 5 five-dollar bills, and 16 one-dollar bills. He put $70 of that money in his bank account. How much money was not put in his bank account?

6. Mrs. Clark has 8 five-dollar bills and 2 ten-dollar bills in her wallet. She has 1 twenty-dollar bill and 12 one-dollar bills in her purse. How much more money does she have in her wallet than in her purse?

Name ______________________________ Date ______________

Solve.

1. Josh had 3 five-dollar bills, 2 ten-dollar bills, and 7 one-dollar bills. He gave Suzy 1 five-dollar bill and 2 one-dollar bills. How much money does Josh have left?

2. Jeremy has 3 one-dollar bills and 1 five-dollar bill. Jessica has 2 ten-dollar bills and 2 five-dollar bills. Sam has 2 ten-dollar bills and 4 five-dollar bills. How much money do they have together?

R (Read the problem carefully.)

Clark has 3 ten-dollar bills and 6 five-dollar bills. He has 2 more ten-dollar bills and 2 more five-dollar bills than Shannon. How much money does Shannon have?

D (Draw a picture.)

W (Write and solve an equation.)

W (Write a statement that matches the story.)

Name ______________________ Date ____________

Write another way to make the same total value.

1. 26 cents     2 dimes 1 nickel 1 penny is 26 cents.	Another way to make 26 cents:
2. 35 cents    3 dimes and 1 nickel make 35 cents.	Another way to make 35 cents:
3. 55 cents    2 quarters and 1 nickel make 55 cents.	Another way to make 55 cents:
4. 75 cents    The total value of 3 quarters is 75 cents.	Another way to make 75 cents:

5. Gretchen has 45 cents to buy a yo-yo. Write two coin combinations she could have paid with that would equal 45 cents.

6. The cashier gave Joshua 1 quarter, 3 dimes, and 1 nickel. Write two other coin combinations that would equal the same amount of change.

7. Alex has 4 quarters. Nicole and Caleb have the same amount of money. Write two other coin combinations that Nicole and Caleb could have.

Name ______________________________ Date ______________

Smith has 88 pennies in his piggy bank. Write two other coin combinations he could have that would equal the same amount.

R (Read the problem carefully.)

Andrew, Brett, and Jay each have 1 dollar in change in their pockets. They each have a different combination of coins. What coins might each boy have in his pocket?

D (Draw a picture.)

W (Write and solve an equation.)

W (Write a statement that matches the story.)

___

___

___

Name ______________________________ Date ____________

1. Kayla showed 30 cents two ways. Circle the way that uses the fewest coins.

a.	b.

What two coins from (a) were changed for one coin in (b)?

____

2. Show 20¢ two ways. Use the fewest possible coins on the right below.

	Fewest coins:

3. Show 35¢ two ways. Use the fewest possible coins on the right below.

	Fewest coins:

4. Show 46¢ two ways. Use the fewest possible coins on the right below.

	Fewest coins:

5. Show 73¢ two ways. Use the fewest possible coins on the right below.

	Fewest coins:

6. Show 85¢ two ways. Use the fewest possible coins on the right below.

	Fewest coins:

7. Kayla gave three ways to make 56¢. Circle the correct ways to make 56¢, and star the way that uses the fewest coins.

   a. 2 quarters and 6 pennies

   b. 5 dimes, 1 nickel, and 1 penny

   c. 4 dimes, 2 nickels, and 1 penny

8. Write a way to make 56¢ that uses the fewest possible coins.

Name ____________________ Date __________

1. Show 36 cents two ways. Use the fewest possible coins on the right below.

	Fewest coins:

2. Show 74 cents two ways. Use the fewest possible coins on the right below.

	Fewest coins:

R (Read the problem carefully.)

Tracy has 85 cents in her change purse. She has 4 coins.

a. Which coins are they?

b. How much more money will Tracy need if she wants to buy a bouncy ball for $1?

D (Draw a picture.)

W (Write and solve an equation.)

W (Write a statement that matches the story.)

a. ______________________________

____

b. ______________________________

____

Name ______________________________ Date ______________

1. Count up using the arrow way to complete each number sentence. Then, use your coins to show your answers are correct.

a. 45¢ + ________ = 100¢

$45 \xrightarrow{+5}$ _____ $\xrightarrow{+\_\_\_}$ 100

b. 15¢ + ________ = 100¢

c. 57¢ + ________ =100¢

d. ________ + 71¢ = 100¢

2. Solve using the arrow way and a number bond.

a. 79¢ + ________ = 100¢

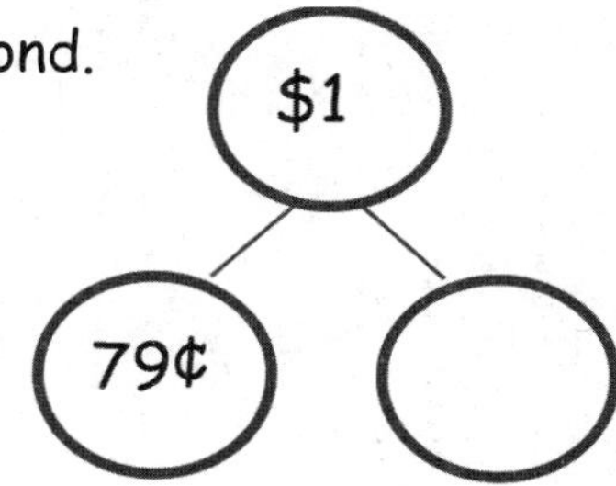

b. 64¢ + ________ = 100¢

c. 100¢ - 30¢ = ________

3. Solve.

a. ________ + 33¢ = 100¢

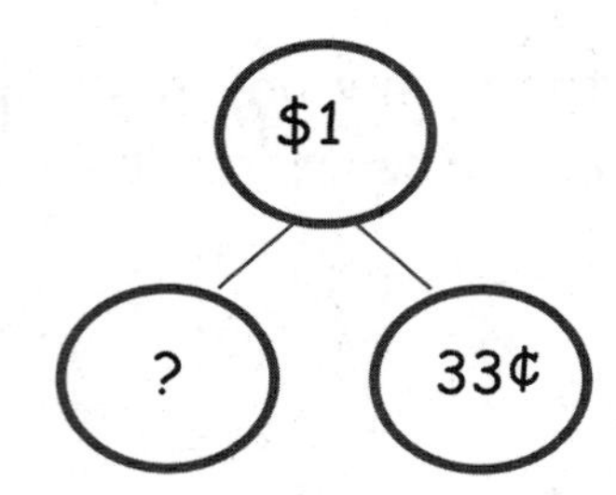

b. 100¢ - 55¢ = ________

c. 100¢ - 28¢ = ________

d. 100¢ - 43¢ = ________

e. 100¢ - 19¢ = ____________

Name ____________________ Date __________

Solve.

1. 100¢ - 46¢ = ________

2. ________ + 64¢ = 100¢

3. ________ + 13 cents = 100 cents

R (Read the problem carefully.)

Richie has 24 cents. How much more money does he need to make $1?

D (Draw a picture.)

W (Write and solve an equation.)

W (Write a statement that matches the story.)

Name ______________________________ Date ______________

Solve using the arrow way, a number bond, or a tape diagram.

1. Jeremy had 80 cents. How much more money does he need to have $1?

2. Abby bought a banana for 35 cents. She gave the cashier $1. How much change did she receive?

3. Joseph spent 75 cents of his dollar at the arcade. How much money does he have left?

4. The notepad Elise wants costs $1. She has 4 dimes and 3 nickels. How much more money does she need to buy the notepad?

5. Dane saved 26 cents on Friday and 35 cents on Monday. How much more money will he need to save to have saved $1?

6. Daniel had exactly $1 in change. He lost 6 dimes and 3 pennies. What coins might he have left?

Name ____________________ Date ____________

Solve using the arrow way, a number bond, or a tape diagram.

Jacob bought a piece of gum for 26 cents and a newspaper for 61 cents. He gave the cashier $1. How much money did he get back?

R (Read the problem carefully.)

Dante had some money in a jar. He puts 8 nickels into the jar. Now he has 100 cents. How much money was in the jar at first?

D (Draw a picture.)

W (Write and solve an equation.)

W (Write a statement that matches the story.)

Name ______________________________ Date ______________

Solve with a tape diagram and number sentence.

1. Josephine has 3 nickels, 4 dimes, and 12 pennies. Her mother gives her 1 coin. Now Josephine has 92 cents. What coin did her mother give her?

2. Christopher has 3 ten-dollar bills, 3 five-dollar bills, and 12 one-dollar bills. Jenny has $19 more than Christopher. How much money does Jenny have?

3. Isaiah started with 2 twenty-dollar bills, 4 ten-dollar bills, 1 five-dollar bill, and 7 one-dollar bills. He spent 73 dollars on clothes. How much money does he have left?

4. Jackie bought a sweater at the store for $42. She had 3 five-dollar bills and 6 one-dollar bills left over. How much money did she have before buying the sweater?

5. Akio found 18 cents in his pocket. He found 6 more coins in his other pocket. Altogether he has 73 cents. What were the 6 coins he found in his other pocket?

6. Mary found 98 cents in her piggy bank. She counted 1 quarter, 8 pennies, 3 dimes, and some nickels. How many nickels did she count?

Name ______________________________ Date ______________

Solve with a tape diagram and number sentence.

Gary went to the store with 4 ten-dollar bills, 3 five-dollar bills, and 7 one-dollar bills. He bought a sweater for $26. What bills did he leave the store with?

Frances is moving the furniture in her bedroom. She wants to move the bookcase to the space between her bed and the wall, but she is not sure it will fit.

What could Frances use as a measurement tool if she doesn't have a ruler? How could she use it?

Show your thinking using pictures, numbers, or words.

Name ______________________________ Date ______________

1. Measure the objects below with an inch tile. Record the measurements in the table provided.

Object	Measurement
Pair of scissors	
Marker	
Pencil	
Eraser	
Length of worksheet	
Width of worksheet	
Length of desk	
Width of desk	

2. Mark and Melissa both measured the same marker with an inch tile but came up with different lengths. Circle the student work that is correct, and explain why you chose that work.

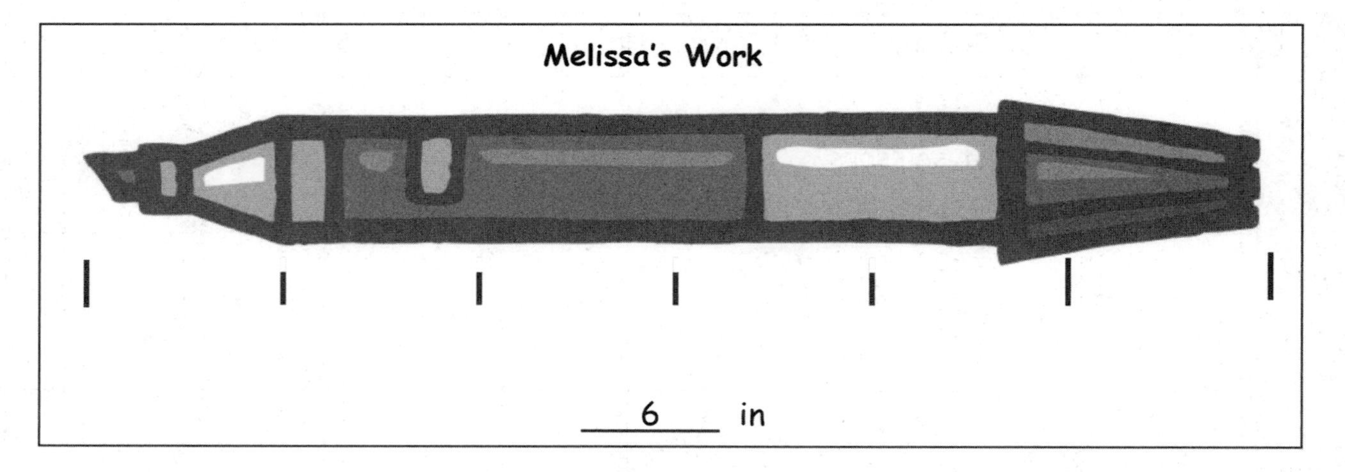

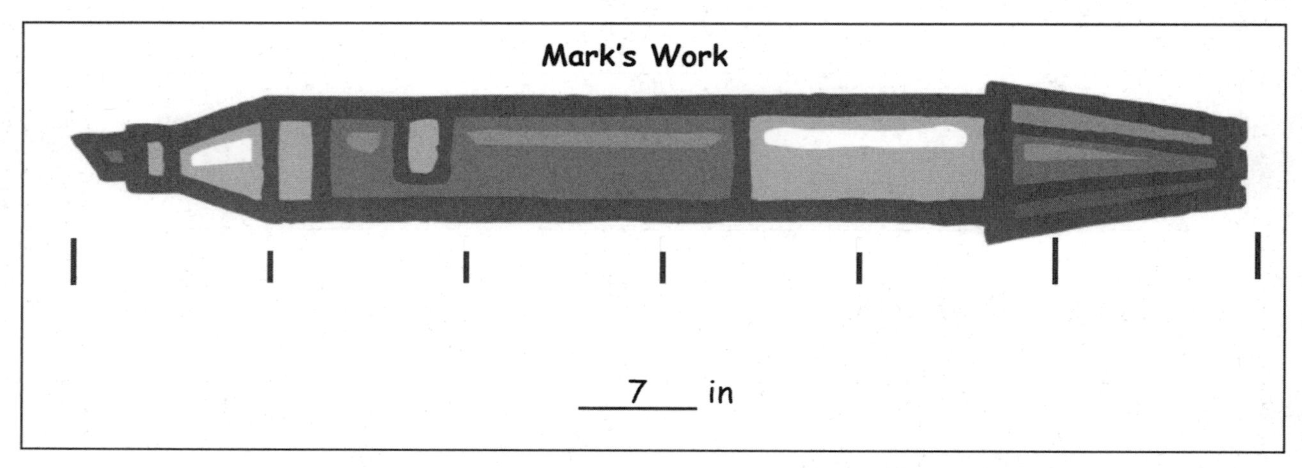

Explanation:

____

____

____

Name ______________________________ Date ______________

Measure the lines below with an inch tile.

Line A ____________________________________

Line A is about ________ inches.

Line B ____

Line B is about ________ inches.

Line C ____

Line C is about ________ inches.

R (Read the problem carefully.)

Edwin and Tina have the same toy truck. Edwin says his is 4 toothpicks long. Tina says hers is 12 lima beans long. How can they both be right?

Use words or pictures to explain how Edwin and Tina can both be right.

D (Draw a picture.)

W (Write and solve an equation.)

W (Write a statement that matches the story.)

Name ______________________________ Date ______________

Use your ruler to measure the length of the objects below in inches. Using your ruler, draw a line that is the same length as each object.

1. a. A pencil is _______ inches.
   b. Draw a line that is the same length as the pencil.

2. a. An eraser is _______ inches.
   b. Draw a line that is the same length as the eraser.

3. a. A crayon is _______ inches.
   b. Draw a line that is the same length as the crayon.

4. a. A marker is _______ inches.
   b. Draw a line that is the same length as the marker.

5. a. What is the longest item that you measured? _______________
   b. How long is the longest item? _______________ inches
   c. How long is the shortest item? _______________ inches
   d. What is the difference in length between the longest and the shortest items? _______________ inches
   e. Draw a line that is the same as the length you found in (d).

6. Measure and label the length of each side of the triangle using your ruler.

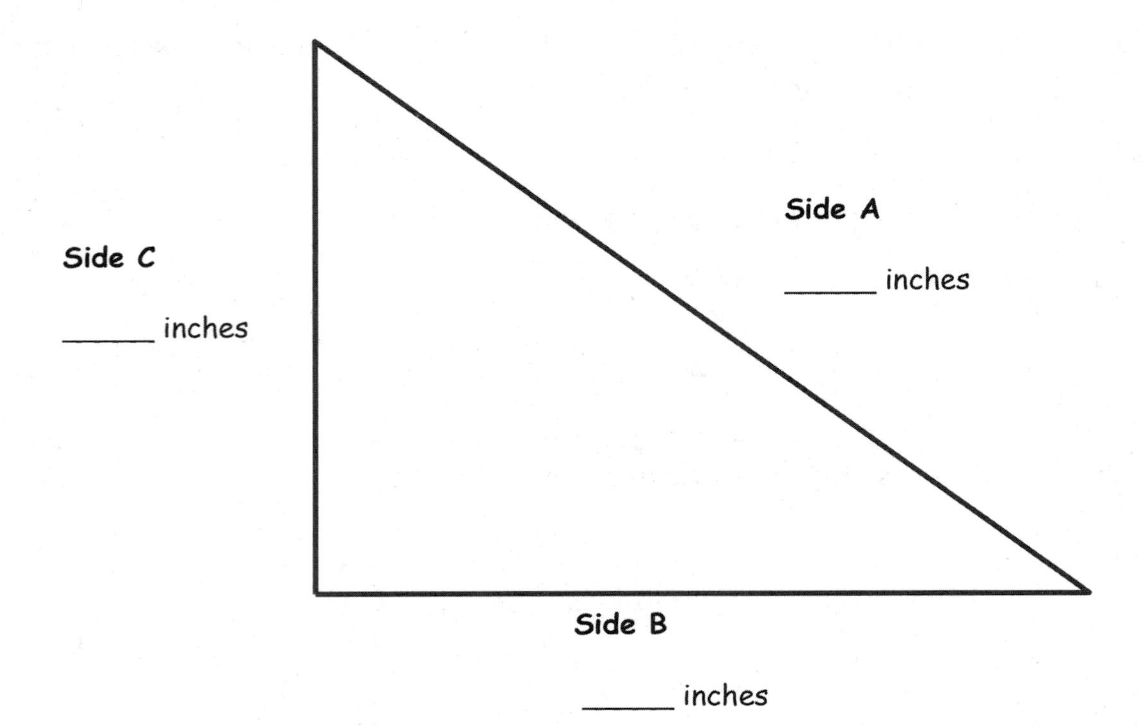

a. Which side is the shortest? Side A Side B Side C

b. What is the length of Side A? __________ inches

c. What is the length of Sides C and B together? __________ inches

d. What is the difference between the shortest and longest sides? __________ inches

7. Solve.

a. _______ inches = 1 foot

b. 5 inches + _______ inches = 1 foot

c. _______ inches + 4 inches = 1 foot

Name ______________________________ Date ______________

Measure and label the sides of the shape below.

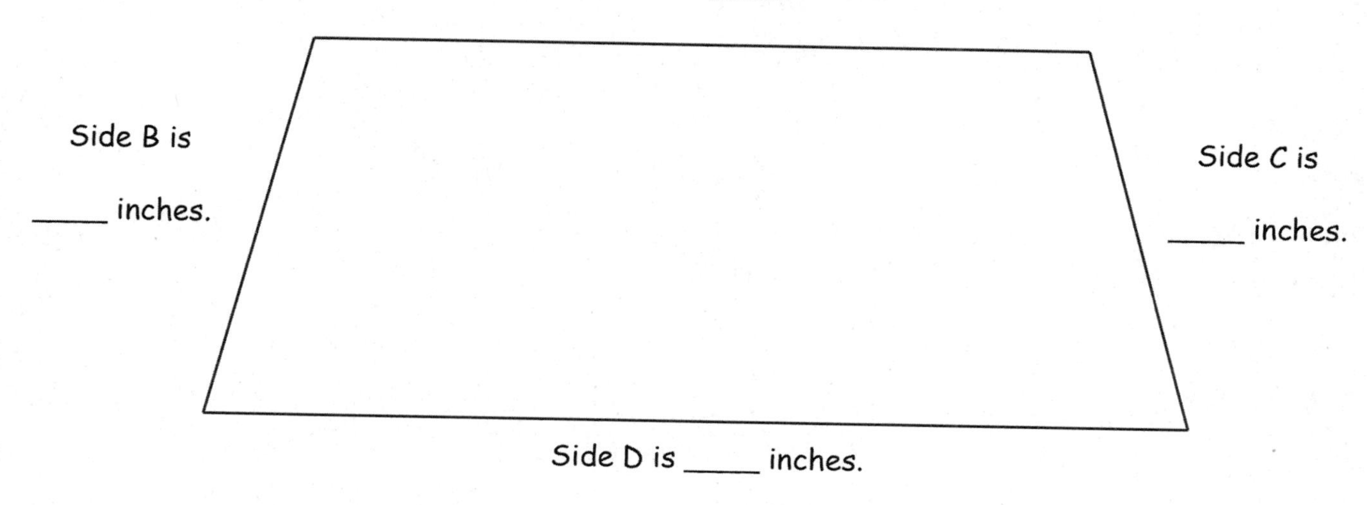

What is the sum of the length of Side B and the length of Side C? __________ inches

Center 1: Measure and Compare Shin Lengths

Choose a measuring unit to measure the shins of everyone in your group. Measure from the top of the foot to the bottom of the knee.

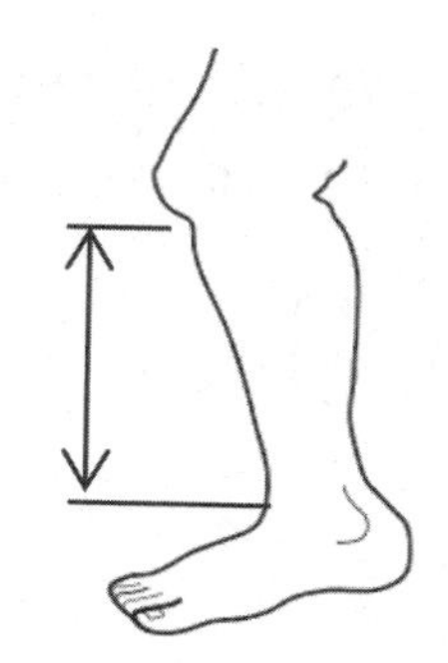

I chose to measure using ____________________.
Record the results in the table below. Include the units.

Name	Length of Shin

What is the difference in length between the longest and shortest shins? Write a number sentence and statement to show the difference between the two lengths.

Center 2: Compare Lengths to a Yardstick

Fill in your estimate for each object using the words *more than, less than, or about the same length as.* Then, measure each object with a yardstick, and record the measurement on the chart.

1. The length of a book is ____________________ the yardstick.
2. The height of the door is ____________________ the yardstick.
3. The length of a student desk is ____________________ the yardstick.

Object	Measurement
Length of book	
Height of door	
Length of student desk	

What is the length of 4 student desks pushed together with no gaps in between? Use the RDW process to solve on the back of this paper.

Center 3: Choose the Units to Measure Objects

Name 4 objects in the classroom. Circle which unit you would use to measure each item, and record the measurement in the chart.

Object	Length of the Object
	inches/feet/yards
	inches/feet/yards
	inches/feet/yards
	inches/feet/yards

Billy measures his pencil. He tells his teacher it is 7 feet long. Use the back of this paper to explain how you know that Billy is incorrect and how he can change his answer to be correct.

Center 4: Find Benchmarks

Look around the room to find 2 or 3 objects for each benchmark length. Write each object in the chart, and record the exact length.

Objects That Are About an **Inch**	Objects That Are About a **Foot**	Objects That Are About a **Yard**
1. ______ inches	1. ______ inches	1. ______ inches
2. ______ inches	2. ______ inches	2. ______ inches
3. ______ inches	3. ______ inches	3. ______ inches

Center 5: Choose a Tool to Measure

Circle the tool used to measure each object. Then, measure and record the length in the chart. Circle the unit.

Object	Measurement Tool	Measurement
Length of the rug	12-inch ruler / yardstick	________ inches/feet
Textbook	12-inch ruler / yardstick	________ inches/feet
Pencil	12-inch ruler / yardstick	________ inches/feet
Length of the chalkboard	12-inch ruler / yardstick	________ inches/feet
Pink eraser	12-inch ruler / yardstick	________ inches/feet

Sera's jump rope is the length of 6 textbooks. On the back of this paper, make a tape diagram to show the length of Sera's jump rope. Then, write a repeated addition sentence using the textbook measurement from the chart to find the length of Sera's jump rope.

Name ______________________________ Date ______________

Circle the unit that would best measure each object.

Marker	inch / foot / yard
Height of a car	inch / foot / yard
Birthday card	inch / foot / yard
Soccer field	inch / foot / yard
Length of a computer screen	inch / foot / yard
Height of a bunk bed	inch / foot / yard

R (Read the problem carefully.)

Benjamin measures his forearm and records the length as 15 inches. Then, he measures his upper arm and realizes it's the same!

a. How long is one of Benjamin's arms?

b. What is the total length of both of Benjamin's arms together?

D (Draw a picture.)

W (Write and solve an equation.)

W (Write a statement that matches the story.)

a. ____

____

b. ____

____

Name ______________________________ Date ______________

Estimate the length of each item by using a mental benchmark. Then, measure the item using feet, inches, or yards.

Item	Mental Benchmark	Estimation	Actual Length
a. Width of the door			
b. Width of the white board or chalkboard			
c. Height of a desk			
d. Length of a desk			
e. Length of a reading book			

Item	Mental Benchmark	Estimation	Actual Length
f. Length of a crayon			
g. Length of the room			
h. Length of a pair of scissors			
i. Length of the window			

Name ________________________________ Date ________________

Estimate the length of each item by using a mental benchmark. Then, measure the item using feet, inches, or yards.

Item	Mental Benchmark	Estimation	Actual Length
a. Length of an eraser			
b. Width of this paper			

Ezra is measuring things in his bedroom. He thinks his bed is about 2 yards long. Is this a reasonable estimate? Explain your answer using pictures, words, or numbers.

Name ______________________ Date ____________

Measure the lines in inches and centimeters. Round the measurements to the nearest inch or centimeter.

1. ________________________

13 cm 5 in

2. ________________________

10 cm 4 in

3. ________________________

15 cm 6 in

4. ________________________

8 cm 3 in

5. a. Did you use more inches or more centimeters when measuring the lines above?

CM

b. Write a sentence to explain why you used more of that unit.

____

____

6. Draw lines with the measurements below.

   a. 3 centimeters long

   b. 3 inches long

7. Thomas and Chris both measured the crayon below but came up with different answers. Explain why both answers are correct.

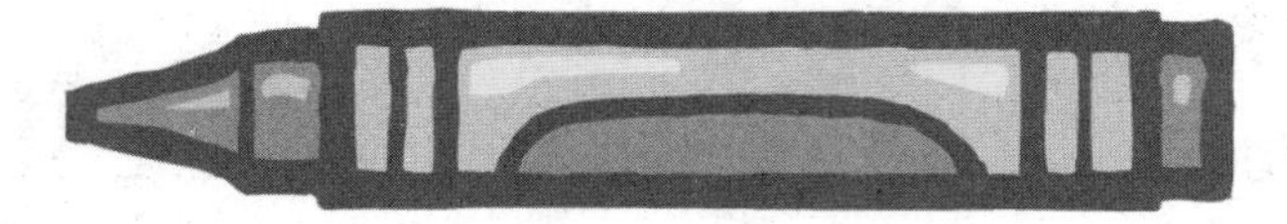

Thomas: 8 cm

Chris: 3 in

Explanation: ______________________________

Name ______________________________ Date ______________

Measure the lines in inches and centimeters. Round the measurements to the nearest inch or centimeter.

1. ____

_______ cm _______ in

2. ____

_______ cm _______ in

R (Read the problem carefully.)

Katia is hanging decorative lights. The strand of lights is 46 feet long. The building wall is 84 feet long. How many more feet of lights does Katia need to buy to equal the length of the wall?

D (Draw a picture.)

W (Write and solve an equation.)

W (Write a statement that matches the story.)

____

____

____

Name ______________________________ Date ______________

Measure each set of lines in inches, and write the length on the line. Complete the comparison sentence.

1. Line A ____

   Line B ________________

   Line A measured about _____ inches. Line B measured about _____ inches.

   Line A is about _____ inches **longer** than Line B.

2. Line C ____________________

   Line D ____

   Line C measured about _____ inches. Line D measured about _____ inches.

   Line C is about _____ inches **shorter** than Line D.

3. Solve the following problems:

a. 32 ft + ________ = 87 ft

b. 68 ft - 29 ft = ________

c. ________ - 43 ft = 18 ft

4. Tammy and Martha both built fences around their properties. Tammy's fence is 54 yards long. Martha's fence is 29 yards longer than Tammy's.

Tammy's Fence	Martha's Fence
54 yards	________ yards

a. How long is Martha's fence? ________ yards

b. What is the total length of both fences? ________ yards

Name ______________________ Date ____________

Measure the set of lines in inches, and write the length on the line. Complete the comparison sentence.

Line A

Line B

Line A measured about _____ inches.

Line B measured about _____ inches.

Line A is about _____ inches **longer/shorter** than Line B.

Name ____________________________ Date ______________

Solve using tape diagrams. Use a symbol for the unknown.

1. Mr. Ramos has knitted 19 inches of a scarf he wants to be 1 yard long. How many more inches of scarf does he need to knit?

2. In the 100-yard race, Jackie has run 76 yards. How many more yards does she have to run?

3. Frankie has a 64-inch piece of rope and another piece that is 18 inches shorter than the first. What is the total length of both ropes?

4. Maria had 96 inches of ribbon. She used 36 inches to wrap a small gift and 48 inches to wrap a larger gift. How much ribbon did she have left?

5. The total length of all three sides of a triangle is 96 feet. The triangle has two sides that are the same length. One of the equal sides measures 40 feet. What is the length of the side that is not equal?

?

6. The length of one side of a square is 4 yards. What is the combined length of all four sides of the square?

Name ______________________________ Date ______________

Solve using a tape diagram. Use a symbol for the unknown.

Jasmine has a jump rope that is 84 inches long. Marie's is 13 inches shorter than Jasmine's. What is the length of Marie's jump rope?

R (Read the problem carefully.)

To ride the Mega Mountain roller coaster, riders must be at least 44 inches tall. Caroline is 57 inches tall. She is 18 inches taller than Addison. How tall is Addison? How many more inches must Addison grow to ride the roller coaster?

D (Draw a picture.)

W (Write and solve an equation.)

W (Write a statement that matches the story.)

Name ______________________________ Date ______________

Find the value of the point on each part of the meter strip marked by a letter. For each number line, one unit is the distance from one hash mark to the next.

1.

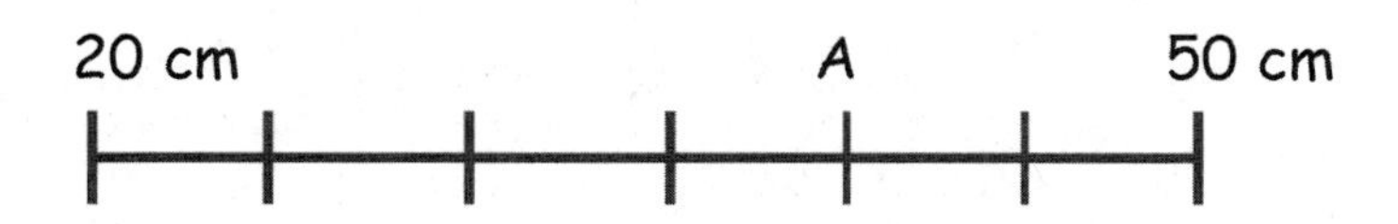

Each unit has a length of __________ centimeters.

A = ____________

2.

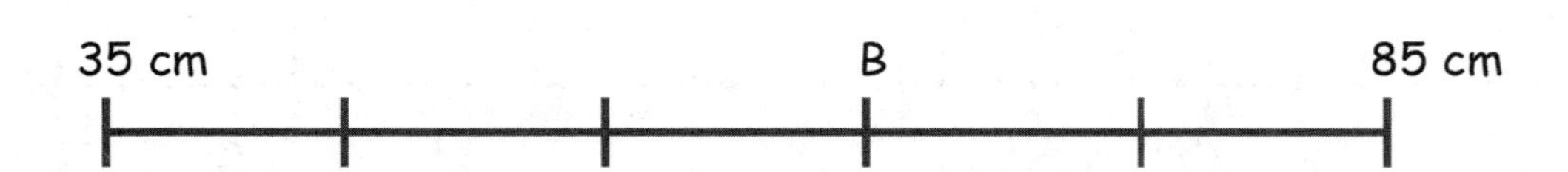

Each unit has a length of __________ centimeters.

B = ____________

3.

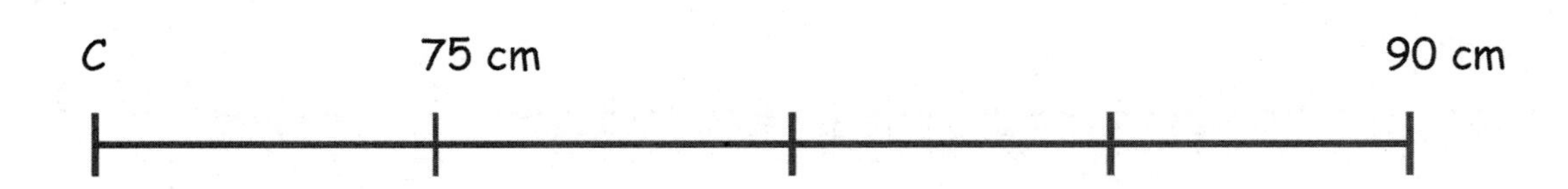

Each unit on the meter strip has a length of __________ centimeters.

C = ____________

4. Each hash mark represents 5 more on the number line.

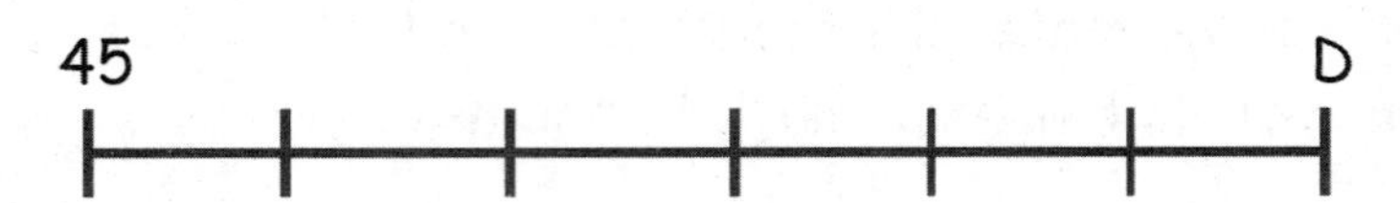

D = ______________

What is the difference between the two endpoints? ___________________.

5. Each hash mark represents 10 more on the number line.

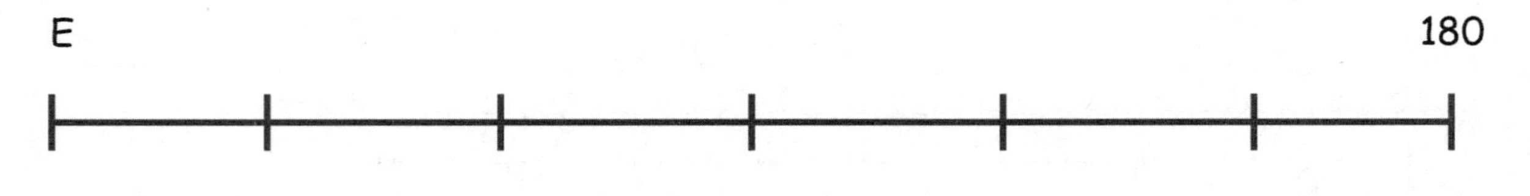

E = ______________

What is the difference between the two endpoints? ___________________.

6. Each hash mark represents 10 more on the number line.

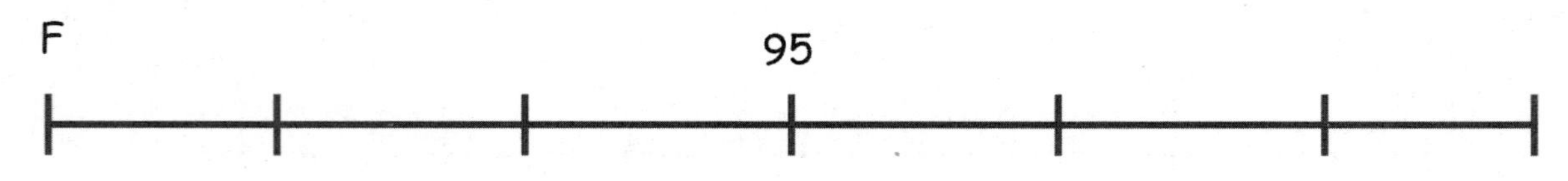

F = ______________

What is the difference between the two endpoints? ______________.

Name ______________________________ Date ______________

Find the value of the point on each number line marked by a letter.

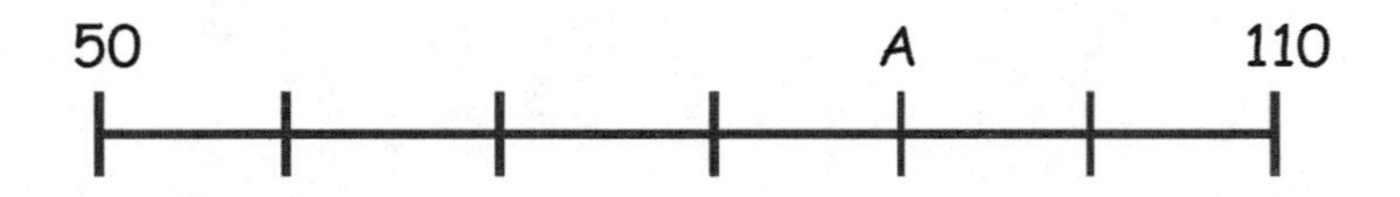

1. Each unit has a length of __________ centimeters.

   A = ______________

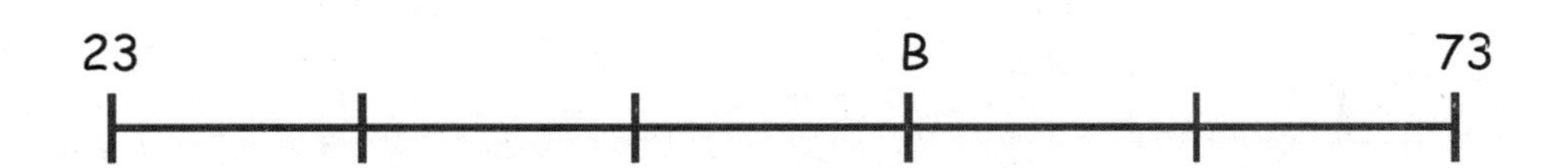

2. What is the difference between the two endpoints? __________.

   B = ______________

R (Read the problem carefully.)

Liza, Cecilia, and Dylan are playing soccer. Liza and Cecilia are 120 feet apart. Dylan is in between them. If Dylan is standing the same distance from both girls, how many feet is Dylan from Liza?

D (Draw a picture.)

W (Write and solve an equation.)

W (Write a statement that matches the story.)

____

____

____

Name ______________________________ Date ______________

1. Each unit length on both number lines is 10 centimeters.
   (Note: Number lines are not drawn to scale.)

   a. Show 30 centimeters more than 65 centimeters on the number line.

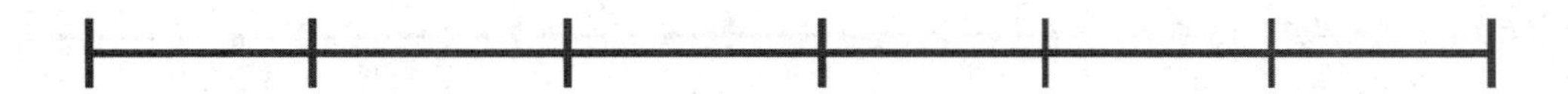

   b. Show 20 centimeters more than 75 centimeters on the number line.

   c. Write an addition sentence to match each number line.

2. Each unit length on both number lines is 5 yards.

   a. Show 25 yards less than 90 yards on the following number line.

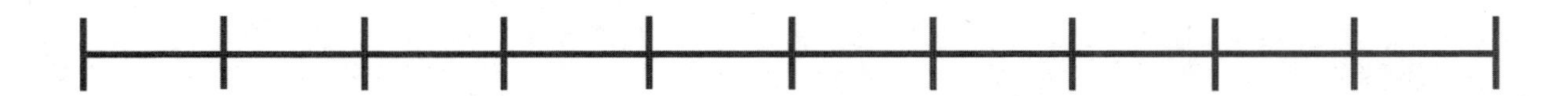

   b. Show 35 yards less than 100 yards on the number line.

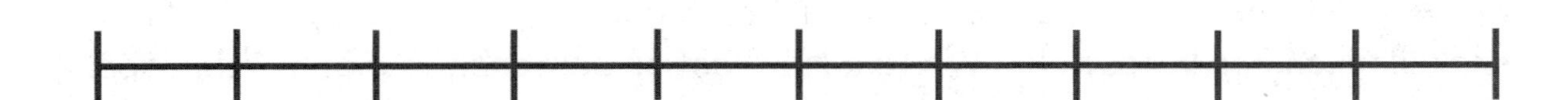

   c. Write a subtraction sentence to match each number line.

3. Vincent's meter strip got cut off at 68 centimeters. To measure the length of his screwdriver, he writes "81 cm - 68 cm." Alicia says it's easier to move the screwdriver over 2 centimeters. What is Alicia's subtraction sentence? Explain why she's correct.

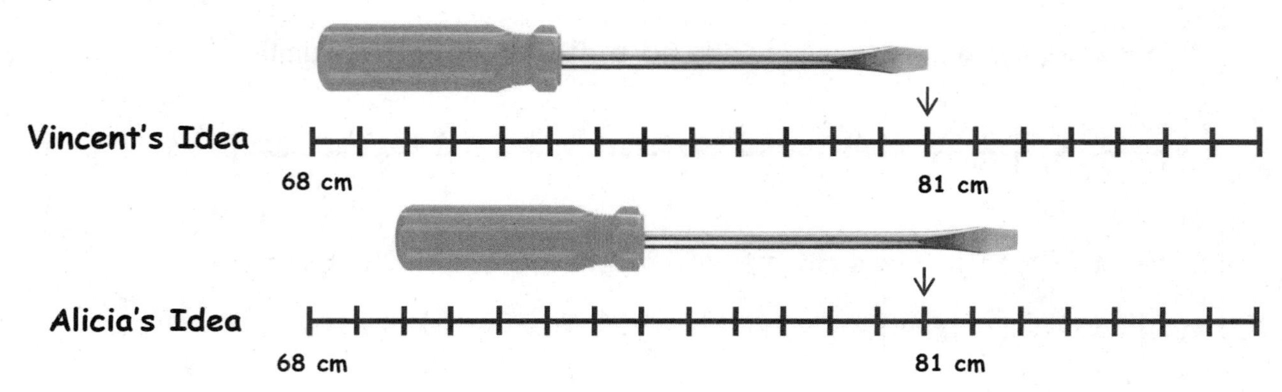

4. A large flute is 71 centimeters long, and a small flute is 29 centimeters long. What is the difference between their lengths?

5. Ingrid measured her garden snake's skin to be 28 inches long using a yardstick but didn't start her measurment at zero. What might be the two endpoints of her snakeskin on her yardstick? Write a subtraction sentence to match your idea.

Name ____________________________ Date ______________

Each unit length on both number lines is 20 centimeters.
(Note: Number lines are not drawn to scale.)

1. Show 20 centimeters more than 25 centimeters on the number line.

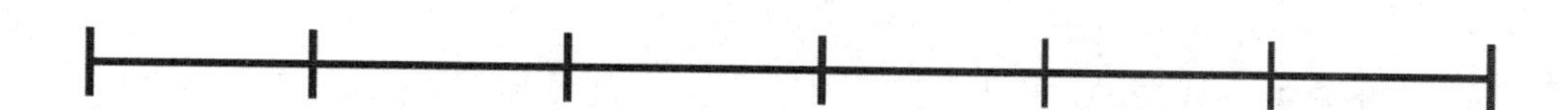

2. Show 40 centimeters less than 45 centimeters on the number line.

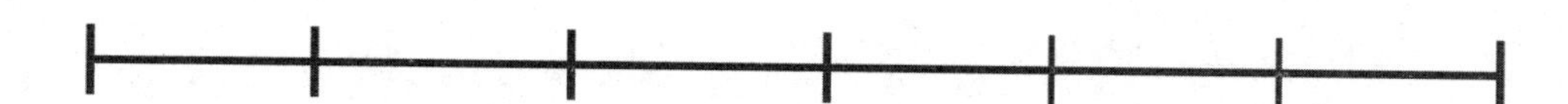

3. Write an addition or a subtraction sentence to match each number line.

Number Line A

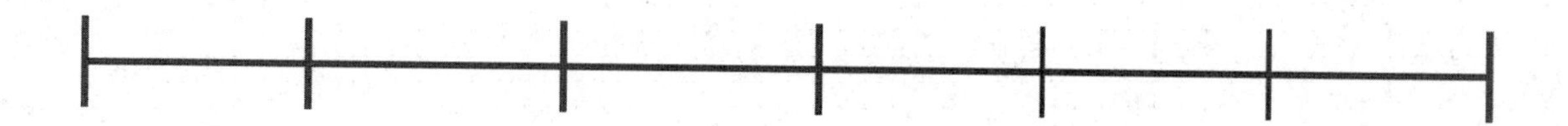

Number Line B

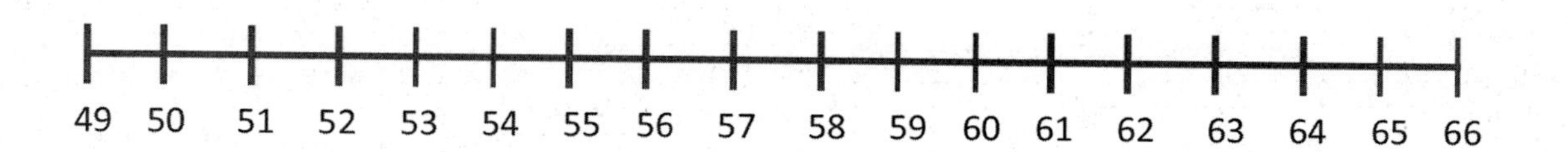

number lines A and B

Name ______________________________ Date ______________

1. Gather and record group data.

   Write your teacher's handspan measurement here: ________

   Measure your handspan, and record the length here: ________

   Measure the handspans of the other people in your group, and write them here. We will be using the data tomorrow.

Name: **Handspan:**

______________________________ ______________

______________________________ ______________

______________________________ ______________

______________________________ ______________

______________________________ ______________

Handspan	Tally of Number of People
3 inches	
4 inches	
5 inches	
6 inches	
7 inches	
8 inches	

What is the most common handspan length? _____

What is the least common handspan length? _____

What do you think the most common handspan length will be for the whole class? Explain why.

____

____

____

____

____

____

2. Record the class data.

Record the class data using tally marks on the table provided.

Handspan	Tally of Number of People
3 inches	
4 inches	
5 inches	
6 inches	
7 inches	
8 inches	

What handspan length is the most common? ______________

What handspan length is the least common? ______________

Ask and answer a comparison question that can be answered using the data above.

Question: ____

____

Answer: ____

____

Name ______________________________ Date ______________

1. Measure the lines below in inches. Record the data using tally marks on the table provided.

   Line A ____

   Line B __________________________

   Line C ______________________________________

   Line D __________________________________

   Line E ______

   Line F _________

   Line G ____

Line Length	Number of Lines
Shorter than 5 inches	
Longer than 5 inches	
Equal to 5 inches	

2. How many more lines are shorter than 5 inches than are equal to 5 inches?

   ____

3. What is the difference between the number of lines that are shorter than 5 inches and the number that are longer than 5 inches? ________________

4. Ask and answer a comparison question that could be answered using the data above.

   Question: ____

   ____

Switch papers with a partner. Have your partner answer your question on the back.

Name ______________________________ Date ______________

1. The lines below have been measured for you. Record the data using tally marks on the table provided, and answer the questions below.

Line A 5 inches

Line B 6 inches

Line C 4 inches

Line D 6 inches

Line E 3 inches

Line Length	Number of Lines
Shorter than 5 inches	
5 inches or longer	

2. If 8 more lines were measured to be longer than 5 inches and 12 more lines were measured to be shorter than 5 inches, how many tallies would be in the chart?

____

R (Read the problem carefully.)

Mike, Dennis, and April all collected coins from a parking lot. When they counted their coins, they had 24 pennies, 15 nickels, 7 dimes, and 2 quarters. They put all the pennies into one cup and the other coins in another. Which cup has more coins? How many more?

D (Draw a picture.)

W (Write and solve an equation.)

W (Write a statement that matches the story.)

____

____

____

Name ______________________________ Date ______________

Use the data in the tables to create a line plot and answer the questions.

1.

Pencil Length (inches)	Number of Pencils
2	I
3	II
4	~~IIII~~ I
5	~~IIII~~ II
6	~~IIII~~ III
7	IIII
8	I

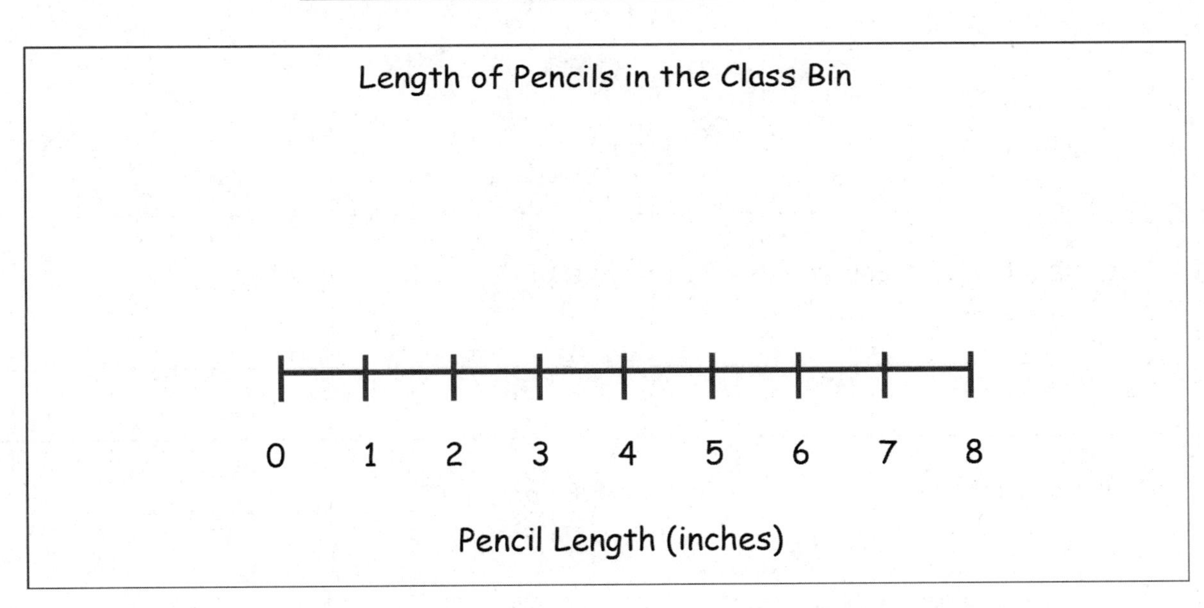

Describe the pattern you see in the line plot:

____

____

2.

Length of Ribbon Scraps (centimeters)	Number of Ribbon Scraps
14	I
16	III
18	~~IIII~~ III
20	~~IIII~~ II
22	~~IIII~~

Scraps of Ribbon in the Arts and Crafts Bin

Line Plot

a. Describe the pattern you see in the line plot.

____

____

b. How many ribbons are 18 centimeters or longer? ______________

c. How many ribbons are 16 centimeters or shorter? ______________

d. Create your own comparison question related to the data.

____

____

Name ______________________________ Date ______________

Use the data in the table to create a line plot.

Length of Crayons in a Class Bin

Crayon Length (inches)	Number of Crayons
1	|||
2	~~||||~~ ||||
3	~~||||~~ ||
4	~~||||~~

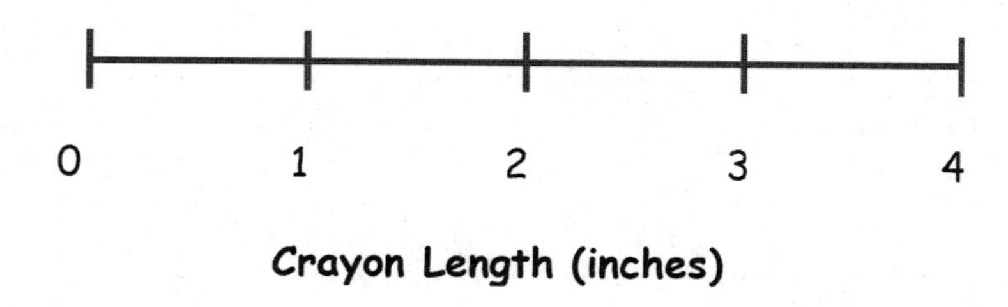

Crayon Length (inches)

R (Read the problem carefully.)

These are the types and numbers of stamps in Shannon's stamp collection.

Type of Stamp	Number of Stamps
Holiday	16
Animal	8
Birthday	9
Famous singers	21

Her friend Michael gives her some flag stamps. If he gives her 7 fewer flag stamps than birthday and animal stamps together, how many flag stamps does she have?

Extension: If the flag stamps are worth 12 cents each, what is the total value of Shannon's flag stamps?

D (Draw a picture.)

W (Write and solve an equation.)

W (Write a statement that matches the story.)

Name ______________________________ Date ______________

Use the data in the chart provided to create a line plot and answer the questions.

1. The chart shows the heights of the second-grade students in Mr. Yin's homeroom.

Height of Second-Grade Students	Number of Students
40 inches	1
41 inches	2
42 inches	2
43 inches	3
44 inches	4
45 inches	4
46 inches	3
47 inches	2
48 inches	1

Title ______________________________

Line Plot

a. What is the difference between the tallest student and the shortest student?

b. How many students are taller than 44 inches? Shorter than 44 inches?

2. The chart shows the length of paper second-grade students used in their art projects.

Length of Paper	Number of Students
3 ft	2
4 ft	11
5 ft	9
6 ft	6

Title ______________________________

Line Plot

a. How many art projects were made? ________

b. What paper length occurred most often? ________

c. If 8 more students used 5 feet of paper and 6 more students used 6 feet of paper, how would it change how the line plot looks?

____

____

d. Draw a conclusion about the data in the line plot.

____

____

Name ______________________________ Date ______________

Answer the questions using the line plot below.

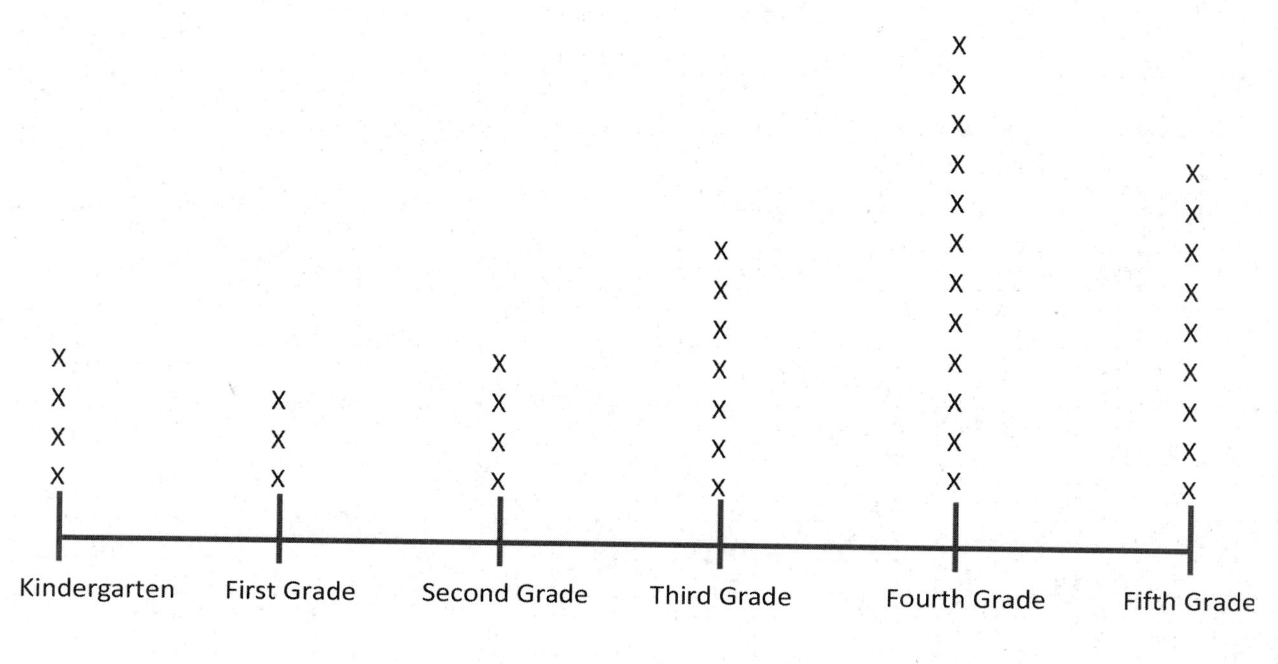

1. How many students went to the baseball game? ________

2. What is the difference between the number of first-grade students and the number of fourth-grade students who went to the baseball game? ________

3. Come up with a possible explanation for why most of the students who attended are in the upper grades.

____

____

R (Read the problem carefully.)

Judy bought an MP3 player and a set of earphones. The earphones cost $9, which is $48 less than the MP3 player. How much change should Judy get back if she gave the cashier a $100 bill?

D (Draw a picture.)

W (Write and solve an equation.)

W (Write a statement that matches the story.)

Name ______________________________ Date ______________

Use the data in the table provided to answer the questions.

1. The table below describes the heights of basketball players and audience members who were polled at a basketball game.

Height (inches)	Number of Participants
25	3
50	4
60	1
68	12
74	18

a. How tall are most of the people who were polled at the basketball game?

____

b. How many people are 60 inches or taller? __________

c. What do you notice about the people who attended the basketball game?

____

d. Why would creating a line plot for these data be difficult?

____

____

e. For these data, a **line plot / table** (circle one) is easier to read because ...

____

____

Use the data in the table provided to create a line plot and answer the questions.

2. The table below describes the length of pencils in Mrs. Richie's classroom in centimeters.

Length (centimeters)	Number of Pencils
12	1
13	4
14	9
15	10
16	10

a. How many pencils were measured? ____________

b. Draw a conclusion as to why most pencils were 15 and 16 cm:

____

____

c. For these data, a **line plot / table** (circle one) is easier to read because...

____

____

Name ______________________________ Date ______________

Use the data in the table provided to create a line plot.

The table below describes the heights of second-grade students on the soccer team.

Height (inches)	Number of Students
35	3
36	4
37	7
38	8
39	6
40	5

Length of Items in Our Pencil Boxes	Number of Items
6 cm	1
7 cm	2
8 cm	4
9 cm	3
10 cm	6
11 cm	4
13 cm	1
16 cm	3
17 cm	2

Temperatures in May	Number of Days
59°	1
60°	3
63°	3
64°	4
65°	7
67°	5
68°	4
69°	3
72°	1

length and temperature tables

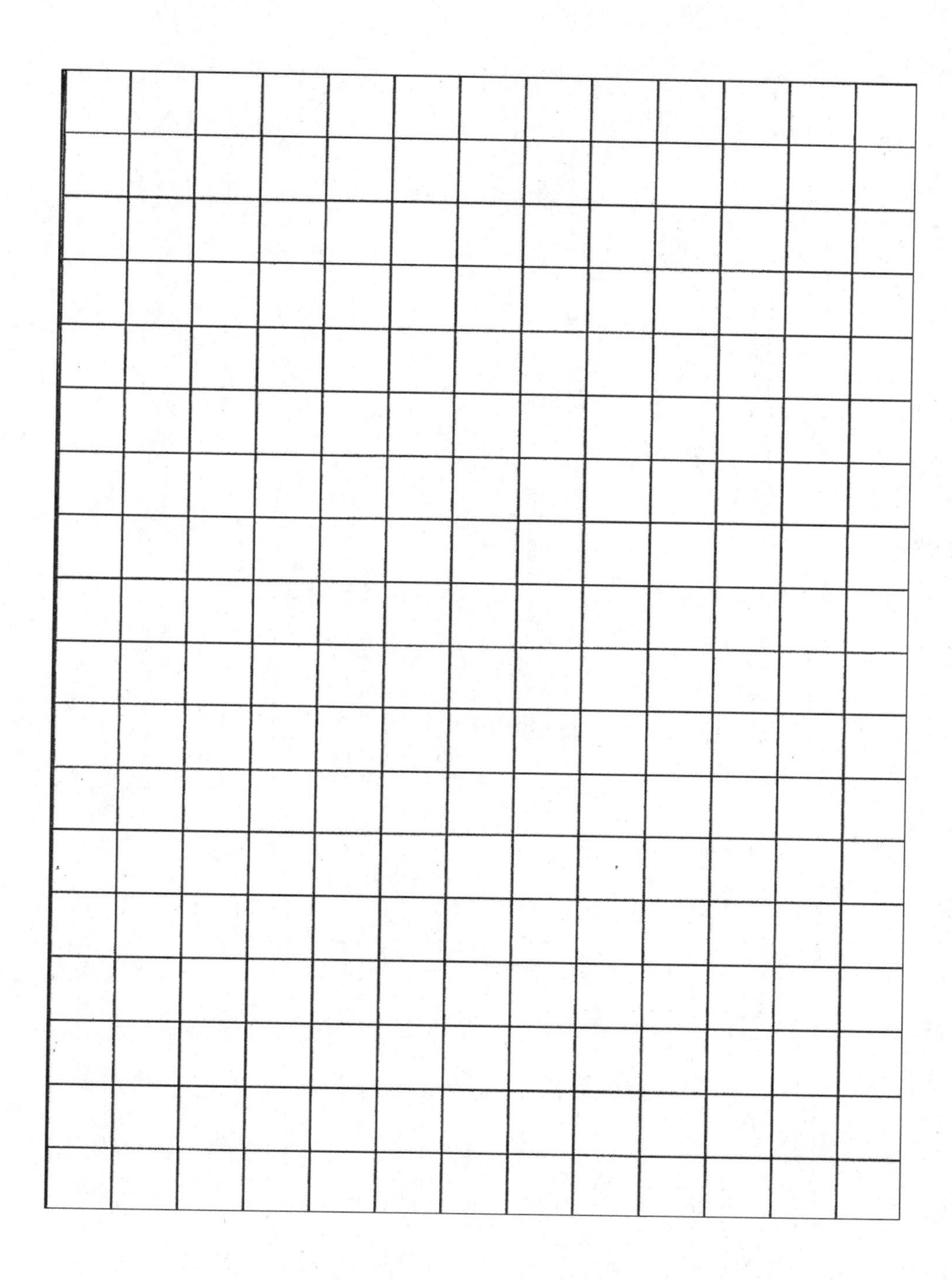

grid paper

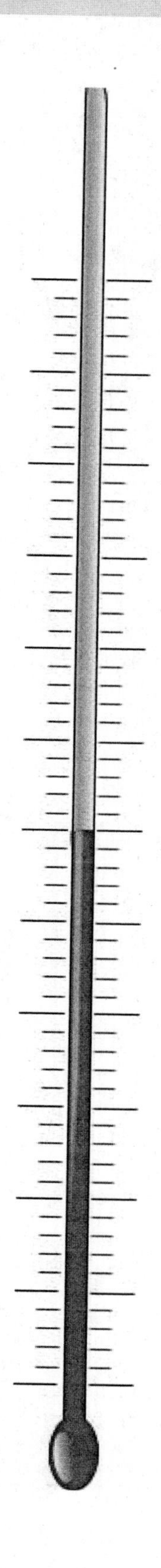

____

thermometer

Credits

Great Minds® has made every effort to obtain permission for the reprinting of all copyrighted material. If any owner of copyrighted material is not acknowledged herein, please contact Great Minds for proper acknowledgment in all future editions and reprints of this module.

- Page 266, Joao Virissimo/Shutterstock.com